Summary

Construction is the largest consumer of natural resources in the UK. More than 90 per cent of non-energy materials extracted in the UK supply the construction industry. This represents, on average, nearly 30 million tonnes per year of primary materials, of which some 214 million tonnes is in the form of aggregates. Concern is growing about the environmental consequences and the long-term sustainability of this resource use. In response, the UK Government is seeking both to reduce the demand for primary aggregates, for example through the Aggregates Levy, and to encourage the use of alternative materials.

This review seeks to identify the potential for using secondary and recycled aggregates, including inert construction and demolition (C&D) waste, in coastal and river engineering schemes. It seeks to reduce the impact of river and coastal construction on natural resources by promoting the use of alternative materials. In addition, it aims to increase the use of alternatives to primary aggregates in the UK (the UK is already a leading user of these materials in Europe), by:

- raising awareness of the potential use of secondary aggregates and recycled/reused materials as aggregates

- reassuring designers and constructors of the appropriateness of using alternatives to primary aggregates

- assisting in overcoming reluctance to the use of alternatives to primary aggregates

- enabling the construction industry to provide more sustainable and cost-effective solutions for river and coastal engineering.

Acknowledgements

Research contractor   This book is the result of CIRIA Research Project 687 "Potential use of alternatives to primary aggregates in coastal and river engineering" carried out under contract to CIRIA by **HR Wallingford Ltd**.

Authors

Alan Brampton BSc PhD
Alan Brampton is technical director of the Coastal & Seabed Group at HR Wallingford. He has 30 years' experience of research and practical studies of waves and beach development, both in the UK and around the world. He has contributed to several major design/guidance publications including the CIRIA Report 153 *Beach management manual* and the ICE *Design and practice guide. Coastal defence*.

Michael Wallis BSc MSc AIEMA
Michael Wallis is a coastal research scientist within the Engineering Systems and Management Group at HR Wallingford and is actively involved in construction materials research and sustainable development in flood and coastal management. He has particular expertise in environmental and integrated coastal zone management.

Elizabeth Holliday BSc MSc
Elizabeth Holliday is a coastal officer at Kent County Council with responsibility for strategic management of the Kent coast. Before this, she was a project manager at CIRIA and has more than six years' experience of managing collaborative research on water management issues, in particular with respect to the coastal and marine environment.

Steering group   Following CIRIA's usual practice, the research project was guided by a steering group, which comprised the following members.

Chairmen   Roger Maddrell and Ben Hamer   Halcrow Group Ltd

Attending members	
John Barritt	WRAP
Mark Buckley	Black & Veatch
Tim Collins	English Nature
Jason Golder	The Crown Estate
Jeremy Henry	Van Oord ACZ
Alan Inder	Hampshire County Council
George Lees	Scottish Natural Heritage
Brian Marker	ODPM
John Mason/David Rathbone	Alan Baxter & Associates
Alastair McNeill	SEPA
Fola Ogunyoye	Posford Haskoning
Chris Vivian	CEFAS
Bill Schlegel	ICE Maritime Board

CIRIA managers   CIRIA's research managers for the project were **Craig Elliott** and **Elizabeth Holliday**.

Project funders   The project was funded by:

CIRIA's Core Programme

DTI Partners in Innovation (PII) programme

The Crown Estate

Van Oord ACZ

Contributors   CIRIA and the authors are grateful for the help given to this project by the funders, the members of the steering group. We also acknowledge the input of the many individuals who were consulted and provided data, particularly those external to the project steering group who provided additional peer-review input:

Charlie Rickard	Independent consulting engineer
John Riby	Scarborough Borough Council
Brian James	Quarry Products Association
Chris Worthy	Tarmac Recycling Ltd
Roger Morris	English Nature
Tony Murray	The Crown Estate
Tim Pinder	RMC UK Ltd
Rod Collins	BRE
Stuart Meakins	Environment Agency
Terry Oakes	Terry Oakes Associates Ltd
Siva Sivaloganathan	Mott MacDonald
Anne Padfield	University of Greenwich
David Brook	ODPM

And all attendees of the project consultation workshop (see Appendix 5).

Responsibilities   The authors of this report are employed by CIRIA. The work reported herein was carried out under a contract jointly funded by CIRIA's Core Programme, the Crown Estate, Van Oord ACZ and the Secretary of State for Trade and Industry, placed on 15 May 2003. Any views expressed here are not necessarily those of the Secrtetary of State for Trade and Industry.

Contents

Figures

Tables and boxes

TABLES

BOXES

Case studies

Glossary

Aggregate	The European Standards for Aggregates provide the following definitions, which should now be regarded as the correct definitions for construction aggregates in the UK and EU.
Aggregate	Granular material used in construction. Aggregate may be natural, manufactured or recycled.
Recycled aggregate	Aggregate resulting from the processing of inorganic material previously used in construction.
Manufactured aggregate	Aggregate of mineral origin resulting from an industrial process involving thermal or other modification.
	In addition to these "standard" terms it is useful here to add two further definitions for aggregates.
Primary aggregate	Construction aggregates produced from crushed rock, and sand and gravel (land and marine won). These aggregates are subject to the Aggregates Levy.
Secondary aggregates	Construction aggregates produced from by-products of industrial processes (manufactured aggregates) such as metallurgical slags, pulverised fuel ash (PFA) and incinerator bottom ash (IBA), plus aggregates produced as by-products from other mineral-extraction processes such that they do not incur the Aggregates Levy, such as china clay sand, and slate aggregates.
Alternative materials	Materials, such as recycled and secondary aggregates, that have not traditionally been considered construction materials.
Armour	Outer protective layer of a sea or river defence usually made of **armour units**.
Armour stone	Large quarried stone used as primary protection against wave attack.
Armour units	Large quarried stone or specially shaped concrete blocks used as primary protection against wave action.
Beach recharge (or renourishment)	Supplementing the natural volume of sediment on a beach, using imported material (also referred to as beach nourishment/feeding).
By-products	Useful materials that arise as a consequence of processes to create primary products.
Cement-bound material/ mixture (CBM)	A mixture of soil or aggregate and Portland cement that generally has a water content compatible with compaction by rolling. After compaction and cement hydration, the mixture hardens to produce a hard, durable and erosion-resistant construction material.
China clay	Commercial term for kaolin, a clay mineral used in the manufacture of whiteware ceramics and in the filling and coating of paper.
China clay sand	Granite sand produced from the processing of waste generated by the china clay industry.
Construction and demolition waste (C&D waste)	Inert waste generated by the construction and/or demolition of buildings and/or civil engineering infrastructure. Materials include concrete, brick, asphalt, unbound aggregates, soil and clay.
Controlled waste	"Household, industrial and commercial waste or any such waste. Such definition includes waste arising from works of demolition, construction and preparatory work thereto" (EPA 1990 and Controlled Waste Regulations 1992).
Core	Material within the defence structure protected by the outer armour or cover layer.
Culvert	A covered conduit for taking a watercourse, drain or sewer under a railway, road or embankment.
Diffuse pollution	Pollution that does not arise from an easily identifiable source (or point source such as a discharge pipe). Usually refers to runoff or leaching from land.

Dioxins	A particularly toxic class of halogenated aromatic compounds, the by-product of the bleaching process used in the manufacture of white paper and the manufacture of other chemicals such as the herbicide Agent Orange and from incomplete incineration of wastes containing chlorine.
Dredging	The excavation and removal of material from the bed of a river, harbour, lake or sea by dredger, dragline or scoop.
Durability	The ability of a material to resist degradation and retain its physical and mechanical properties.
Ecosystem	The plants, animals and micro-organisms that live in a defined space and the physical environment in which they live.
Embankment	Earth structure, often built for flood protection.
Environment	Surroundings in which an organism operates, including air, water, land, natural resources, flora, fauna, humans and their interrelation.
Embodied energy	The quantity of energy required by all of the activities associated with a production process including the acquisition of primary material, transportation, manufacturing and handling.
Estuary	A semi-enclosed body of water in which seawater is substantially diluted with freshwater entering from land drainage.
Gabion	A generally rectangular box or mattress made of wire mesh and filled with stone.
Geotextile	A synthetic fabric used as a filter or a separation layer (can be woven or unwoven).
Groyne	A structure, generally perpendicular to the shoreline or riverbank, built to control the movement of beach material or to arrest erosion.
Hoggin	Sieved or screened gravel.
Hydraulically bound material/mixture (HBM)	Includes **CBM** and thus soil cement and roller-compacted concrete, but also describes similar products based on the cementitious properties of lime, pulverised fuel ash and hydraulic slag such as ground granulated blast-furnace slag.
Leachate	The liquid generated after a solid is subjected to a leachant.
Lithology	The physical characteristics of a rock or sediment, including colour, composition and texture.
Nutrient	A substance providing nourishment for living organisms (such as nitrogen and phosphorous).
Permeability	The ease with which water will pass into and through the pores in rock, soil, sand or gravel.
Permeameter	An instrument for measuring permeability.
Pollution	The introduction by man into the environment of substances or energy liable to cause hazards to human health, harm to living resources and ecological systems, damage to structures or amenity, or interference with legitimate uses of the environment.
Pozzolanic	Ability to harden when reacted with lime and water to give a cementitious product.
Pyrolysis	Decomposition of rubber by heat in an oxygen-free atmosphere.
Primary aggregates	See **Aggregate** above
Primary materials	Materials extracted from virgin natural reserves.
Porosity	Ratio of voids of a material to its total volume.
Quarry	A site from which natural rock is extracted.
Ramsar site	A site designated by government as a Wetland of International Importance (Ramsar site) under the Convention on Wetlands of International Importance Especially as Waterfowl Habitat (the Ramsar Convention, 1973).

Recycling	System of collecting, sorting and reprocessing old material into usable raw materials.
Recycled aggregates	See **Aggregates** above.
Resource	A concentration of materials from which extraction of a commodity may be possible.
Reuse	The employment of an article or item once again for its original purpose, or for a different purpose, without prior processing to change its physical or chemical characteristics.
Revetment	One or more layers of stone, concrete or other material used to protect the sloping surface of an embankment, natural coast or shoreline against erosion.
Rip-rap	Widely graded rock armour.
Rubble-mound structure	A mound of randomly shaped and randomly placed stone.
Salmonid waters	Waters in which game fish (such as salmon, trout, grayling and whitefish) are found.
Scalpins	Small stones used for drainage in excavations, and as hardcore.
Scour	Erosion of bed or beach material close to a structure due to wave or river action.
Secondary aggregates	See **Aggregates** above.
Site of Special Scientific Interest (SSSI)	An area of land or water notified under the Wildlife and Countryside Act 1981 (as amended) as being of geological or nature conservation importance, in the opinion of the Countryside Council for Wales, English Nature or Scottish Natural Heritage.
Special Area of Conservation (SAC)	Established under the EC Habitats Directive (92/43/EEC), implemented in the UK by The Conservation (Natural Habitats etc) Regulations 1994 and The Conservation (Natural Habitats etc) (Northern Ireland) Regulations 1995. The sites are significant in habitat type and species, and are considered in greatest need of conservation at a European level. All UK SACs are based on SSSIs, but may cover several separate but related sites.
Special Protection Area (SPA)	The Directive on Conservation of Wild Birds (79/409/EEC) allows for the designation of areas specifically for bird species to prevent deliberative capture, killing and disturbance of certain endangered species as well as the destruction or damage to eggs and nests. (Designated sites are also subject to the provisions of the Conservation (Natural Habitats etc) Regulations 1994.)
Spoil	Soil or rock or other material arising from excavation, dredging or other ground engineering work.
Toe	The lowest part of a coastal or river defence structure or riverbank.
Under-layer	Granular layer beneath armour or cover layer which may serve as a filter and/or as a separating layer. May be replaced or augmented by a geotextile.
Waste	Waste is something that "the producer or holder discards or intends to or is required to discard" (Waste Management Licensing Regulations 1994). For a substance or object to be waste it must:
	fall into one of the categories set out in Part II of Schedule 4 to the Regulations; and
	• be discarded or disposed of by the holder; or
	• be intended to be discarded or disposed of by the holder; or
	• be required to be discarded or disposed of by the holder
	(from DoE Circular 11/94).

Abbreviations

ASLQ   Area of Special Landscape Quality
ASR   alkali-silica reaction
BFS   blast-furnace slag
BOF   basic oxygen furnace (slag)
BOS   basic oxygen steelmaking
BPOE   best practicable environmental option
BRE   Building Research Establishment
BS   British Standard
C&D   construction and demolition
CBM   concrete-bound material
CCW   Countryside Council for Wales
CDEW   construction and demolition excavation waste
CEN   European Standards Committee
CEFAS   Centre for Environment, Fisheries & Aquaculture Science
CFMP   catchment flood management plan
Defra   Department for Environment, Food and Rural Affairs
DoE   Department of the Environment
DTI   Department of Trade and Industry
D50   particle size for which 50 per cent of the material is finer
EA   Environment Agency
EAF   electric arc furnace (slag)
EC   European Commission
EMP   estuary management plan
EN   English Nature
EPA   Environmental Protection Act
EU   European Union
FAS   flood alleviation scheme
FBA   furnace bottom ash
FEPA   Food and Environmental Protection Act
GGBS   ground granulated blast-furnace slag
HBM   hydraulically bound material
IBA   municipal solid waste incinerator bottom ash
IDB   internal drainage board
kt   thousand tonnes
MSW   municipal solid waste
mt   million tonnes
mt/a   million tonnes per annum
ODPM   Office of the Deputy Prime Minister
PFA   pulverised fuel ash
pH   acidity/alkalinity
PR   public relations
RCC   roller-compacted concrete
RTB   railway track ballast
SAC   Special Area of Conservation
SEPA   Scottish Environmental Protection Agency
SHW   Specification for Highway Works
SNH   Scottish Natural Heritage
SMP   shoreline management plan
SPA   Special Protection Area
SSSI   Site of Special Scientific Interest
TRL   Transport Research Laboratory
WFD   Water Framework Directive
WML   waste management licence
WRAP   Waste and Resources Action Programme

1   Background and use of the book

1.1   BACKGROUND TO THE STUDY

Construction is the largest consumer of natural resources in the UK. More than 90 per cent of the non-energy minerals that are extracted in the UK supply the construction industry. This represents, on average, nearly 300 million tonnes a year (mt/a) of primary materials (Smith, Kersey and Griffiths, 2002), the majority of which (some 214 mt/a) is in the form of aggregates. If, as expected, the UK's demand for aggregates increases by 1 per cent per cent per annum, then by 2012 an extra 20 mt of aggregates will be needed annually (see <http://www.aggregain.org.uk>). Concern is growing about the environmental consequences and the long-term sustainability of providing this large amount of construction material.

Reflecting this, in April 2002 the UK Government introduced an Aggregates Levy as an environmental tax on the commercial exploitation of aggregates in the UK. Presently set at £1.60 per tonne, the main objectives of this levy are to reduce the demand for primary aggregates and encourage the use of alternative materials (see <http://www.hmce.gov.uk/business/othertaxes/agg-levy.htm>).

This latter objective is particularly relevant to the present project. The UK is already a leading user of these materials in Europe and has established large and successful markets for alternatives to primary aggregates. In England alone, some 50 mt/a are derived from recycled or secondary sources, as identified in previous research (see <http://www.aggregain.org.uk>). Increasing the use of alternative (eg recycled) construction materials is a potentially more sustainable way to meet future demands.

Several definitions of aggregate are used in this book. The following standard terms from the European Standards for Aggregates should now be regarded as the correct definitions for construction aggregates in the UK and EU:

- **aggregate** – granular material used in construction; it may be natural, manufactured or recycled

- **recycled aggregate** – aggregate resulting from the processing of inorganic material previously used in construction

- **manufactured aggregate** – aggregate of mineral origin resulting from an industrial process involving thermal or other modification.

To these standard terms it is useful to add two further definitions:

- **primary aggregate** – construction aggregates produced from crushed rock and sand and gravel (land- and marine-won). These aggregates are subject to the Aggregates Levy

- **secondary aggregates** – construction aggregates produced from by-products of industrial processes (manufactured aggregates) such as metallurgical slags, pulverised fuel ash (PFA,) and incinerator bottom ash (IBA), plus aggregates produced as by-products from other mineral extraction processes such that they do not incur the Aggregates Levy, such as china clay sand and slate aggregates.

In addition to being a major consumer of natural resources, the construction industry is also one of the largest generators of waste in the UK, producing approximately 150 millions tonnes of waste each year (Smith, Kersey & Griffiths, 2002). This, coupled with limited available landfill space and the implementation of the EU Landfill Directive, influenced the UK Government's introduction of the Landfill Tax (see <http://www.hmce.gov.uk/business/othertaxes/landfill-tax.htm>) and waste strategy (see <http://www.defra.gov.uk/environment/waste/strategy/cm4693>) in an attempt to secure changes to behaviour and to meet new waste targets. Some inert construction and demolition (C&D) materials are still going into landfill. Increased recycling of such materials would further reduce the demand for primary aggregates for new construction projects.

In the specific case of coastal and river engineering activities, the UK uses each year about 1 million tonnes of armour-stone and about 2 million tonnes of largely sea-won aggregates, at a value well in excess of £100 million. At present, this usage consists almost entirely of primary materials, such as marine-dredged sand and gravel for beach recharge schemes and high-quality rock, predominantly from coastal quarries. Because of recent severe river flooding, and the predicted acceleration of sea level rise, it is expected that the increase in the demand for materials will be proportionately greater in this sector of civil engineering than for general construction.

It is therefore important that coastal and river engineers address their resource usage and reduce consumption wherever possible. The Environment Agency has introduced targets to encourage the use of alternatives to primary aggregates to this end. In addition to reducing the basic resource demands, there may be a further potential advantage to the natural environment in reducing the impacts (CO_2 emissions, embodied energy etc) associated with the extraction, processing and transport of primary materials to construction sites.

Furthermore, the water sector of civil engineering can provide a crucial pathfinder role in promoting new applications of alternatives to primary materials. There are three reasons for this:

- the sensitivity of the water environment, limiting the levels of allowable contaminants

- the harsh conditions, especially on coasts, requiring use of strong and durable materials

- the need to keep scheme costs low, given the limited funding available from the Exchequer.

Successful applications in this sector would help promote further applications across the entire construction industry. This will contribute to a general reduction in the demand for primary resources in construction and in the disposal of materials from construction and demolition to landfill. It will also lead to a specific improvement in the sustainability of river and coastal engineering, reducing their demands for primary aggregates.

1.2 OBJECTIVES AND SCOPE OF STUDY

The strategic, long-term objective of this study is to reduce the impact of river and coastal construction on natural resources by promoting the use of alternative materials in place of primary aggregate and other materials.

The project's aims are to:

- raise awareness of the potential use of secondary aggregate and recycled/reused materials as aggregates

- reassure designers and constructors of the appropriateness of using alternatives to primary aggregates

- help address the barriers to the use of alternatives to primary aggregates in a strategic and co-ordinated way, and hence

- enable the construction industry to provide more sustainable and cost-effective solutions for river and coastal engineering;

To assist in achieving the above aims, this study had the following subsidiary targets:

- to identify river and coastal engineering works that could use alternative materials

- to review the availability and suitability of recycled and secondary materials available that could replace primary aggregates

- to identify any barriers to the use of these materials, and how they might be overcome

- to present guidelines for the possible use of alternative materials, with case histories describing where they have been used

- to identify where further development or research could extend the use of such materials.

This project is principally a scoping exercise, reviewing the potential for the use of secondary and recycled aggregates, including inert C&D waste, in coastal and river engineering schemes. The target audience for this publication is the construction industry, demolition contractors, coastal and river engineers, and environmental and minerals regulators.

The initial phase of the study concentrated on reviewing the types of river and coastal engineering schemes that are carried out in the UK and the quantities, availability and characteristics of secondary and recycled aggregates.

In both of these areas, attention was first focused on the large-volume ends of the spectrum, for example materials arising from demolition, and beach recharge projects. Beach recharge projects over the whole of the UK involve the use of 1–3 mt of material each year. Overcoming prejudices and barriers to replacing primary aggregates in such high-profile schemes would have a major effect on the wider use of alternative materials. However, it also became clear that local, small-scale uses of recycled or secondary aggregates were already under way, providing valuable examples for wider dissemination. This project therefore considers a wide range of possible applications where alternatives to primary aggregates could be considered.

The project has involved widespread consultation and discussions with experienced practitioners in both river and coastal engineering and in waste recycling. Examples of the deliberate and considered use of alternatives to primary aggregates in coastal and river engineering have been sought, within the UK and worldwide. Where possible, past schemes have been collated and presented as case histories, to demonstrate that potential does exist for further and perhaps more ambitious applications.

However, this review together with a workshop held as part of the project, also revealed some barriers to the widespread use of secondary and recycled materials in this form of civil engineering.

These included concerns about:

- potential risks to water quality and to the wider natural environment (eg by pollution)

- the quality and consistency of alternative materials, and hence their durability

- the possible effects on amenity and aesthetic value of rivers and coasts

- the availability and costs of alternative materials, and the continuity of their supply.

This project sought to separate perceptions from the reality of these issues, and the publication indicates ways in which the remaining obstacles to using alternative materials may be overcome.

During the project, particularly during the consideration of barriers, the need for additional research work has been identified that cannot be satisfied within the scope or duration of this project. This is particularly the case with regard to:

- assisting with the development of policy that will help to take this application forward

- evaluating and revising specifications

- implementing and monitoring pilot projects.

The book therefore makes recommendations on the further work that is needed to overcome prejudices and demonstrate the successful uses of alternatives to the use of primary aggregates in coastal and river engineering projects.

1.3   LAYOUT OF THE BOOK

Section 1.4 of this book briefly explains the project methodology. Thereafter, it has been laid out to match the five short-term targets described in Section 1.2.

Chapter 2 reviews the types of river and coastal engineering schemes around the UK and includes some provisional indications of the types and quantities of materials that are used in their execution.

Chapter 3 provides a description and analysis of the secondary aggregates and recycled inert C&D waste that are available in the UK. This is largely based on information supplied by WRAP's AggRegain initiative (see <http://www.aggregain.org.uk>), supplemented by an initial view on the likely usefulness of each type of material identified for this particular project.

Chapter 4 deals with the barriers to the use of secondary aggregates and of recycled inert C&D materials in river and coastal works. It discusses the environmental and also the practical engineering issues that arise in these situations.

Chapter 5 summarises guidelines for the uses of alternatives to primary aggregates, based on the information gathered during this study. Where possible, this guidance is supported by case histories.

The final chapter (6) sets out recommendations for the future research and development that is likely to be needed to increase the usage of secondary and recycled aggregates, including inert C&D waste. Particular emphasis is placed on practical measures such as trial schemes, rather than on further desk studies or reviews, and on the dissemination of information on such schemes to help convince other engineers that reducing primary aggregates is not only possible but beneficial. These recommendations are based on the findings of consultation with the project steering group and key stakeholders, discussion at the project workshop and through a peer review of this book.

1.4 PROJECT METHODOLOGY

The study began with a review of information both on alternative materials and on past examples of river and coastal engineering schemes that have made use of them. This was carried out through literature and Internet searches and by consultations with the project steering group members and with a wide range of organisations in the UK and overseas.

The review was subdivided into three main topic areas:

- potential alternative construction materials (restricted to England, Wales and Scotland)

- examples of the use of these materials in coastal and river environments (accidental or as engineering works)

- potential barriers to the usage of such materials (concern about pollution or aesthetics, for example).

The last of these three topics was the subject of a project workshop. This was organised because of the importance attached to identifying, and where possible overcoming, any barriers to using these alternative materials.

As a result of the work carried out by WRAP under the AggRegain initiative, much was already known about the types, quantities and availability of secondary aggregates, but there was rather less detail on the types of inert C&D waste that might be usable. When investigating these materials, the project ranged further than seeking recycled granular materials, and considered the larger elements of demolition rubble such as kerbstones, concrete railway sleepers and the like. This project also considered some secondary aggregates, such as china clay sand, slate aggregate, dredged material and scrap tyres.

The second part of the project collated, analysed and reported on the information gathered, drew up preliminary guidelines for the various groups of organisations with an interest in processing, selling, licensing and using alternative materials in coastal and river engineering.

Finally, the main conclusions from the project were summarised, and recommendations drawn up for further work aimed at reducing the use of primary aggregates in coastal and river engineering.

2 Coastal and river engineering requirements

Flood defences are essential in some areas to reduce the risk of flooding to human life. They also reduce risks to property, the loss of which can be both distressing and costly. Potential consequences of climate change for the UK include extreme weather conditions that could lead to more frequent and more severe floods. Storm damage may also be more severe, causing increased erosion of coastal areas and higher maintenance costs for flood defences. This means that sustainability of coastal and river engineering is essential.

In England and Wales alone, some 1.7 million residential properties, 140 000 industrial and commercial properties and extensive areas of agricultural land with an economic value in excess of £200 billion are at risk from flooding and coastal erosion. The current annual average damage arising from flooding and coastal erosion is around £400 million; without investment in mitigation measures this could rise over a period of 50 years to as much as £2 billion a year (Environment Agency, 2001).

In a survey conducted for the Environment Agency's 2001 Flood Defence Investment Strategy for England (Halcrow Maritime, 2001), regional Environment Agency (EA) offices gave their spend in the 1999/2000 financial period for maintenance and replacement of river and sea defences. For river and related defences, maintenance costs amounted to more than £35 million, and replacement costs of just under £122 million. For sea and tidal defences, maintenance costs equated to more than £17 million and replacement costs of slightly less than £97 million. The annual spending on maintenance and replacement of coastal and river defences therefore amounted to more than £271 million (see Appendix 1 for a breakdown of replacement and maintenance costs). This excludes the expenditure incurred by local authorities, for example on coast protection schemes, and on schemes carried out in other parts of the UK.

With many defences in poor condition and the threat of greater dilapidation and inadequacy in the face of more severe storms and higher sea levels, this cost is likely to rise.

2.1 COASTAL ENGINEERING SCHEMES

Coastal engineering works present a range of challenges that can affect the methods of working employed, the type of structures built and the materials used. These challenges can be classified as physical, environmental and economic.

Physical challenges include:

- the aggressive marine environment, eg wave forces, abrasion, corrosion
- poor access and lack of storage and working space during construction
- tidal level changes necessitating unusual working hours, wet/dry construction methods.

The environmental challenges include:

- the need to avoid or minimise physical disturbance or damage to habitats (above and below sea level)
- the potential for damage by spills, leaching of contaminants, washing-out of fine sediments etc
- the need to limit disturbance to local residents from traffic, noise and vibration, especially at night
- concerns about amenity, recreation, landscape and tourism, for example on aesthetics and beach quality.

The main economic challenge arises because, in the UK, most coastal engineering projects are carried out either to reduce erosion of the land or reduce marine flooding risks. These schemes are publicly funded, so there is always a need to keep costs low.

This need for parsimony has stimulated considerable inventiveness and improvisation in the design of coastal defences over the years. In the absence of "ideal" materials, particularly hard rock, coastal engineers have substituted such alternatives as recycled concrete rubble, steel mill slag and used kerbstones in order to keep costs down. For this book, the emphasis has been placed on coast protection and beach management schemes, as these are the most common types of coastal engineering works in the UK and are the ones most likely to be influenced by any national incentives and targets for reduction in the use of primary aggregates.

The logical starting-point for the consideration of alternatives to primary aggregates or other virgin materials in coastal management schemes is to review the typical elements of such schemes. The sketch below shows the most common of these elements for a shoreline backed by cliffs or coastal slopes.

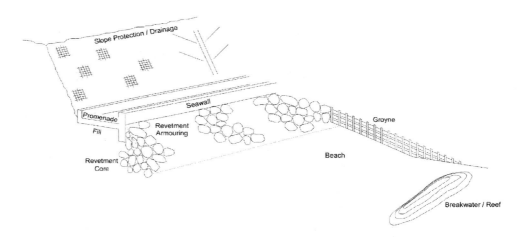

Figure 2.1   *Example elements of a coastal engineering scheme*

Note that flood embankments are also widely used to prevent marine inundation of a low-lying coastal hinterland, for example along many estuaries, but are similar to those built alongside rivers. These structures are discussed in Section 2.2.1.

2.1.1   Beach sediments, recharge and recycling

Beaches act as an important component of the defences against flooding or erosion of the hinterland, and in some areas provide the only defence against these threats. While the continued presence of a beach has often been assumed as a part of the design of coastal defences, it is increasingly recognised that maintaining adequate beach levels is just as important as the appropriate design and construction of, for example, a seawall. Beaches are also an important environmental asset, both for their amenity value to human society and for the habitats they provide for a wide range of plants and animals.

Taken together with the greater size and versatility of marine aggregate dredging vessels, these factors have led to beach recharge becoming an increasingly important component of coastal defence construction over the past 30 years or so. At present, the average quantity of sediments involved in such operations is of the order of 2 mt/a; this is predominantly sand although there have been a number of major shingle beach recharge schemes in recent years as well. The total cost of such schemes is typically in excess of £10 million a year.

Beaches around the UK typically consist of sand or gravel particles in a range of sizes. The largest particles are usually found on the beach surface, often near the beach crest, while smaller particles are found below the surface. The result of this natural sorting of the sediments is that the surface layers of a beach are typically far more permeable than the underlying finer sediments, which can form an impermeable "core". While some beaches have a large proportion of carbonate sand, most beach sediments are of much harder material, often of flint, chert or quartzite. This hardness means that the rate of attrition of individual sand grains or pebbles is low, despite the harsh conditions experienced in the surf zone. The constant wave action, however, does result in particles that generally are rounded in shape.

When carrying out a beach recharge scheme, the normal approach is to seek sources of sediment that are a good match to the "native" sediments. Usually, the sediment added to beaches during such operations is collected from licensed dredging areas on the seabed. These areas have been developed to provide sand and gravel for the construction industry and lie much farther offshore and in deeper water than the sources used in other countries where beach recharge is carried out. A beach recharge operation therefore often uses large quantities of primary aggregates both initially and then periodically afterwards to make good the inevitable losses brought about by the continued erosion of the coastline by the sea.

The main licensed dredging areas are well situated relative to the coastlines most likely to need recharge schemes – off East Anglia and off the central southern coast of England. Despite sometimes needing to transport sand and gravel for more than 100 km between source and recharge site, the cost of such schemes remains competitive with other forms of coastal defence. The costs of providing the necessary sand and gravel, however, are not added to by imposition of the Aggregates Levy since this does not apply to marine dredged primary aggregate when used for beach recharge schemes.

Designers of beach recharge schemes typically specify that the sediment to be delivered is as coarse as, or coarser than, that found on the surface of the beach, in order to minimise losses along the coastline or offshore. While this is often feasible if the beach is of sand, serious problems and expense can be encountered in obtaining large shingle. There are sometimes further restrictions, or at least preferences, for sediments that are rounded, of similar colour to the "native" beach material, and free of any significant amounts of fine-grained particles (eg less than 0.1 mm diameter) to avoid creating turbidity in the nearshore zone during and after placement.

It might be thought, therefore, that these very stringent requirements would make the use of alternative aggregates impractical. However, many of the apparently natural beaches in the UK – particularly in Cornwall, such as in Carlyon and St Ives Bays – are largely the result of dumping waste from quarrying, mining and even from steel-making industries. In other words, they are made up from secondary aggregates. Despite this provenance, they provide both a valuable coastal defence function and an environmental asset for the area.

To date, there have apparently been no deliberate beach recharge schemes carried out using "secondary aggregates" as defined in the present project. Several schemes have been carried out using sediments dredged from navigation channels, however (for example, at Bournemouth, Montrose and most recently at Sandbanks, near Poole). Another instructive scheme was the restoration of a beach at Sconser on the Isle of Skye, using stones (primary aggregate) from a nearby quarry that normally provides roadstone. Here the lithology of the beach sediments and the quarry stone were virtually identical. To avoid altering the character of the beach, the extra stone was placed in a trench at the back of the active beach, with the excavated beach material being moved seawards to replace that which had been lost. This type of technique could equally well be applied for some secondary aggregates such as china clay sand or slate aggregate to reduce the use of primary aggregates in a beach recharge scheme.

More recently, some more inventive beach management schemes have utilised alternatives to marine-dredged primary aggregates, including materials derived from excavation works, recycled concrete rubble from demolition and even bales of old tyres. Where such alternative materials can be placed in the "core" of a beach this adds to the overall volume of it, without the need for the supply of coarse grained primary aggregates.

In several areas of the UK, principally in Sussex and Kent, depleted beaches are recharged using sediments "borrowed" from beaches elsewhere that naturally gain material. These operations are variously described as "by-passing", "back-passing" (in the USA) or "recycling" (in the UK) and normally do not involve the use of any primary aggregates. That said, however, these operations could provide an opportunity to introduce suitable, secondary aggregates or inert C&D waste on a "little and often" basis, thereby adding to the total volume of beach sediments.

2.1.2   Dune management

As an adjunct to managing sandy beaches, it is often necessary to carry out schemes to improve dunes at the beach crest. Such works are usually of low cost, involving fencing, installing boardwalks, planting dune-binding grasses and occasionally covering bare sand surfaces that have been eroded by wind action with "thatching", ie laying down brushwood cuttings, netting or using bituminous sprays.

The main cost in such operations is the labour. The materials used are usually inexpensive, for example old railway sleepers, thinnings from forestry operations and secondhand fishing nets or old fencing from farms. The amount of primary aggregates used in such operations is very small, although in the USA glass beads were used at one site to increase the sand volume in some eroded dunes. Similar localised dune recharge could perhaps be considered in some locations in the UK. The major constraint, apart from cost, would be concerns about the possible adverse effects on plants and animals, since coastal sand dunes are often designated (eg as an SSSI or an SAC) because of their biological importance.

2.1.3   Beach control structures

The direct management of beach sediment volumes by recharge and recycling, as described in Section 2.1.1, is typically accompanied by the installation of beach control structures such as groynes, sills or offshore (or detached) breakwaters. These are designed to reduce the movement of sediments along the coastline, or more rarely offshore, by altering the currents and waves that cause that movement, and/or by directly blocking the transport pathway.

The annual costs of building, maintaining or replacing such structures are considerable, but not easily separable from the costs of beach recharge and recycling operations (see Appendix 1). Groynes around the UK coastline have traditionally been made of tropical hardwood, and this material still predominates. Concrete and sheet-steel piles have also been used in some areas.

An increasing number of rock groynes are being built, either simple linear structures or larger and with a more complicated plan shape (eg T-head or Y-head types), which are designed to affect the incoming waves as well as the longshore currents that the waves and tides produce.

Offshore breakwaters and sills around the UK have all been armoured with large rock (eg at Elmer in Sussex) or with concrete armour units (eg at New Brighton), and some of these structures have smaller-diameter materials within their core. Examples of various types of beach control structures can be found in CIRIA Report 153 *Beach management manual* (Simm, 1996).

Past uses of alternatives to such heavy construction elements have been rather limited. Some use has been made of secondhand wooden railway sleepers and sections of rail to build low-cost groynes in more sheltered areas, and occasionally kerbstones or recycled concrete rubble has been used, usually retained in gabion baskets, to form part or complete groynes or training walls near estuary mouths.

It seems likely that potential future usage of alternative materials, in an unbound form, will be as fill for large rock groynes or breakwaters, or as under-layer or even surface armouring in sheltered locations. Additionally, secondary aggregates could be used in concrete armour units for more exposed locations. The requirements and usage for materials in these structures is very similar to that for rock revetments, as discussed later in Section 2.1.6.

It is worth noting here that several artificial reefs have been installed in relatively deep water (15 m at low tide or more) to provide an ecological niche for marine life, and there is discussion at present about installing multi-purpose reefs in shallower water. These reefs could also provide ecological niche habitats, as well as altering wave conditions along a coastline (and hence the beach plan shape) and producing good wave conditions for surfing. In deeper-water areas, reefs have been built of concrete with PFA as a constituent, which reduces the need for primary aggregates. In shallower-water areas, the risk of damage is greater and more care is needed in the choice of materials. One construction technique being promoted is the use of large geotextile bags filled with granular materials. Some secondary aggregates may offer an attractive alternative to the use of primary aggregates for this purpose. Possible effects on the environment of a bag breaking open need to be considered, however, which might limit the range of particle sizes of the aggregates used to fill the bags.

2.1.4   Saltmarsh management

In estuaries, the mudflats in the inter-tidal zone, and the vegetated saltmarsh areas at and above the high-water mark, play a similar role in coastal processes to the beaches and dunes of sandy shorelines on open coasts.

Saltmarshes, in particular, dissipate wave action and hence limit the forces on flood embankments built around the edges of the estuary. In such situations, where saltmarshes have been reduced in width by erosion, many simple earth embankments have been damaged, leading to the requirement for higher embankments with at least their front faces armoured with concrete blocks or rock revetments. The original, local excavation of soil (usually clay) for their construction was often carried out without a minerals extraction licence, although such excavations are sometimes considered to be in the same category as excavation for sand and gravel for other construction uses. The addition of concrete or rock to armour such embankments clearly does involve the use of primary aggregates (and this topic is returned to later in Section 2.1.6).

One way to avoid the expense of strengthening flood embankments in estuaries is to reverse the erosion of saltmarshes, by directly managing them. Various methods have been used in attempts to achieve this aim, for example the construction of brushwood groynes or "polders", designed to trap and retain mud in front of the marshes, or of breakwaters of various types to reduce wave action along the seaward margins of the marshes. Because of the very small particle sizes involved, there are great difficulties in carrying out recharge schemes in an equivalent way to those on sand beaches. However, increasing use is being made of the materials arising from nearby navigation dredging, both to create low-crest banks (called artificial "cheniers") of sand and gravel to protect saltmarsh areas, and to recharge the mudflat or saltmarsh areas directly.

Should such methods prove successful, there may be a role for secondary aggregates in similar schemes in areas where there is no nearby navigation dredging to provide the required materials. An example might be using crushed, recycled concrete rubble as a substitute for dredged gravel to form low-crest island breakwaters in front of the saltmarshes. As with sand dunes, the potential effects on the biology of saltmarshes and mudflats would need to be examined very carefully. Such areas are generally even more important habitats than sand dunes, and many estuaries in the UK have been designated as of international importance on nature conservation grounds.

2.1.5   Concrete seawalls

Vertical or steeply sloping seawalls are the most common type of coastal defence along urban seafronts in the UK. Concrete is used, as it enables the construction of defences that are inexpensive, compact and quick to build, yet able to withstand the impact forces from the sea. These large impact forces together with the continual problem of abrasion, especially behind shingle beaches, mean that very-high-quality concrete is needed to provide an acceptable structure. Design codes for the standards of concrete to be used in marine applications, and more particularly for seawalls, already exist and include the use of some alternative aggregates (see Section 3.4). Note that simple flood walls, which do not have to deal with large wave forces, are discussed in Section 2.2.2.

Some secondary aggregates are already used for building seawalls. For example, blast-furnace slag has been used as a facing material on seawalls in Lincolnshire. More commonly, secondary aggregates are used in manufacturing the concrete, particularly blast-furnace slag, which can be used either as an aggregate or, if ground and granulated, as a cement replacement.

Many seawalls were built as a thin skin of concrete over a core of granular material, often beach sediments but sometimes over domestic and other refuse. The use of alternative aggregates for fill either behind new seawalls or to replace material lost from beneath existing walls, has the potential to offer further savings in primary aggregates (and indeed preserving the existing stocks of beach sediment). This application is considered further in the discussion of flood embankments in Section 2.2.1.

2.1.6　Rock revetments

As with groynes, concrete seawalls are increasingly being replaced or augmented by the use of rock in the form of sloping revetments. While sometimes regarded as less aesthetically pleasing and potentially hazardous to beach users, the increased roughness and permeability of such revetments acts to reduce wave overtopping and to stem or even reverse beach scour at the toe of the structure. This, together with the ease of delivery (ie by sea) and construction, has made them increasingly popular and cost-effective as coastal defences. Annual expenditure in the UK on this type of structure is probably already in excess of £1 million and is likely to increase as many concrete seawalls reach the end of their design lives. Together with the construction of rock groynes and offshore breakwaters, 0.5–1.0 mt of rock is being used in coastal defence projects each year, much of it imported from overseas.

Rock revetments can range from modest fillets at the base of a seawall to much larger structures that provide the primary or only defence against the sea. In locations where wave action can be severe, the surface layer of the revetment will typically comprise large rocks, or sometimes concrete armour units with masses ranging from 3 t to 10 t and occasionally more. Much the same applies to large breakwaters, for example those built to protect coastal harbours.

There are important technical considerations in choosing the materials used in such revetments. It has often proved difficult to find rock that does not fracture either during the original construction phase, or shortly afterwards as storm waves break over the structure. Once this occurs, the broken remains can be removed from the structure, and/or hurled against it, causing further damage. In addition, especially on shingle coastlines, the abrasion caused by the continual motion of the beach sediments can cause further problems.

There is little scope for direct replacement of such large elements using alternative materials. The maximum unit weight of possible replacements to primary aggregates identified in this study is about 250 kg (the equivalent of concrete railway sleepers). However, in more sheltered locations, or below the upper layers of primary armour units, such revetments can contain considerably smaller units, ranging down to small stones. Recycled concrete rubble from old seawalls, for example, could be used to form the core or an under-layer within new rock revetments. Further, secondary aggregates could be used in the manufacture of concrete armour units.

2.1.7　Other types of seawall

While the previous two categories cover most seawall structures found along the UK coastline, there are others, usually in sheltered or estuarine waters. These include vertical sheet-steel piling walls, wooden bulkhead walls or "breastworks" and gabion or crib-work structures, in which smaller rocks or similar are constrained within a framework of wire, steel, wood etc. Many of these types of wall are filled at their rear with granular material, in the same way as concrete walls, and often with very similar materials. The opportunities for using alternatives to primary aggregate are much the same as for concrete walls (see Section 2.1.5) and embankments (see Section 2.2.1).

There are no obvious methods for using alternative aggregates to replace sheet-steel piling (eg along the banks of tidal rivers) or for timber bulkheads or breastworks, although these do not use much primary aggregate. However, some schemes have been built using demolition waste (ie recycled aggregate) where there is a need for low-cost forms of construction. These materials can be used in gabions, such as at Primrose Wharf on the Thames (Case Study A3.13), or in crib-work structures (Case Study A3.5) to produce low seawalls in front of eroding cliffs such as on the Norfolk coast.

2.1.8 Promenades

The intense wave impact forces on the face of a seawall are in marked contrast to the less severe forces that are experienced on the promenade at the seawall crest, or on the crest of a rock revetment or embankment. Nevertheless there is often a need for some sort of protection to these areas, to deal with the damage that can be caused by over-topping waves and often to provide an access for plant to carry out maintenance works, for emergency vehicles as well as a walkway for pedestrians.

In very exposed conditions, for example a harbour breakwater, the surface has to be of concrete or masonry, to withstand the forces; more typically however, a relatively thin skin of concrete or asphalt/roadstone is sufficient. In these circumstances it can be possible to use alternatives to primary aggregates, for example by crushing and reusing concrete from an old section of seawall, bound in a bituminous mix to provide a surface to the promenade. The Environment Agency used this technique for a seawall reconstruction scheme at St Mary's Bay in Kent.

2.1.9 Cliff and coastal slopes – surface drainage

Surface drainage is an important element of most schemes undertaken to reduce the rate of recession of "soft rock" cliffs, particularly those of glacial till which are found, for example along the coastlines of East Anglia and the central south coast of England. A variety of granular fill materials can be used with the open surface drains, and alternatives to primary aggregates are potentially useful in this role. Recycled aggregate (broken concrete rubble) has, for example, been used for this purpose in cliff drains just to the east of Cromer, Norfolk.

2.1.10 Cliff and coastal slopes – slope protection

Where spray from waves, rainfall, weathering or freeze-thaw processes are a factor in cliff recession, then some type of surface protection may be applied. Examples are netting, geotextile cloths, gabion baskets, articulated concrete mattresses and a thin concrete skin laid over the graded face of the cliff or slope. Although little expenditure on this type of work is needed annually, there is potential for replacing primary aggregates with C&D waste, or possibly secondary aggregates, in:

- gabion baskets
- articulated concrete mattresses
- concrete skin laid over the graded face of the cliff or slope.

2.2 RIVER ENGINEERING SCHEMES

The discussion of river engineering schemes provided here, is intended to be applicable to works on rivers, streams, canals, tidal inlets, and estuaries. Generally speaking, river engineering schemes have a great deal in common with those along the coastline, with the exception that the wave forces that they encounter are much smaller. Except in the few cases where the flow velocities are very high, the impact pressures on such river engineering schemes are considerably less. The forces due to hydrostatic pressure differences and the shear stresses and abrasion brought about by currents, however, remain. As previously, the starting-point for this section is to review existing engineering schemes along rivers and the use of primary aggregates in such schemes. Flood defence works provide a good coverage of the schemes that are likely to be encountered along channels, rivers and tidal inlets in the UK, and the most common elements in such schemes are identified in Figure 2.2, and discussed in more detail in the remainder of this chapter.

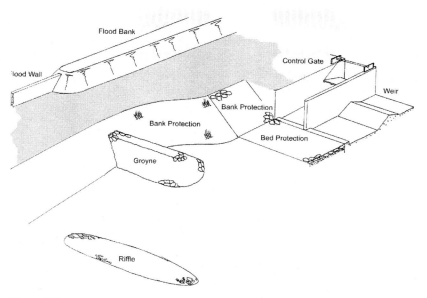

Figure 2.2   *Example elements of a river engineering scheme*

2.2.1   Flood embankments

Based on the cost summaries provided in Appendix 1, expenditure on flood embankments greatly outweighs that on any other type of coastal protection and flood defence works in the UK. In this book, no differentiation has been made between such embankments alongside rivers, estuaries, tidal inlets or the coast. In some situations, however, the wet side of the embankment may require protection in the form of revetment, while in others grass will suffice. Taking all such embankments into consideration, the estimated annual maintenance and replacement costs in the UK are about £25 million and £95 million respectively.

Internal structure

Flood embankments have traditionally been built close to the edge of watercourses, usually using earth dug locally. If this material is pervious, then it is often necessary to provide the embankment with an impervious core, or a deep layer of impervious material over its wet side, to prevent piping or seepage. The crest height of the core, or of the impermeable surface layer, needs to be greater than that of the selected flood level of the water (eg a 1:200 flood level) to be defended against. If, however, the excavated material is impervious, for example clay, then there is no need for any such internal structure to the embankment.

Such local excavations may or may not be treated as "minerals extraction" by the local planning authority, whose advice should be sought at an early stage. However, there appears to be no logical difference, from the minerals planning viewpoint, between this practice and excavating sand or gravel from a quarry and using it to produce a concrete flood wall. Therefore, avoiding such local excavations would be a further way of reducing the usage of primary aggregates.

A disadvantage of such embankments is that their cross-sectional area increases as the square of the crest height, and the resulting weight of the structure, can lead to the geotechnical failure of the ground on which it is built, especially during periods of high water level. This possibility becomes greater when it is necessary to add a revetment on the face of the embankment to prevent surface damage by currents or waves. Such failure of the underlying ground will typically cause a sliding and rotation of the embankment, possibly to the extent of it breaching.

There are several alternatives to excavating materials locally, and to the use of primary aggregates, to form an embankment. Examples include recycled aggregates such as excavation waste or concrete rubble from construction/demolition projects. There may be positive advantages over primary aggregates in using lightweight secondary aggregates, such as ash, and waste tyres bound into bales, allowing an increase in the dimensions of a flood embankment without it becoming too heavy and causing geotechnical failure of the ground beneath it.

Slope protection

The need for protection on the wet side of a flood embankment is dictated by the forces imposed by the currents and waves that it experiences and the strength of the materials from which it is made. As noted previously, there are many earth flood embankments built around the edges of estuaries that have become exposed to large waves as a saltmarsh has eroded, and so have required substantial slope protection on their seaward face. Embankments along rivers or upstream tidal reaches are less likely to require such protection, although concentrations of currents during spate flows might make this necessary, for example adjacent to walls or bridge abutments, or where the plan-shape has led to a strong curvature in the line of the embankment.

Many slope protection measures have been used on embankments, but these are generally much the same as those used for the protection of river and channel banks, as in Section 2.2.3.

2.2.2 Flood walls

In urban areas, where there is insufficient space to build an earth flood embankment, the usual defence structure is a flood wall. The annual replacement cost for such structures in the UK is estimated to be about £10 million. Flood walls are typically vertical, of thin cross-section and built using concrete or sheet-steel piles, although masonry and timber are also used. To cope with the hydrostatic loading on their wet side during flood events, these walls require good support, usually provided by their foundations. A typical concrete wall has an inverted T shape, while sheet-steel pile walls are of the cantilever I type, driven into the ground to provide good support. Sections of such flood walls, for example across roads and paths or at the top of a slipway down a river bank, are often demountable, ie they can be removed until a flood warning is issued and then slotted back into place until the hazard has passed. The crucial design issues, as for embankments, are structural stability and the prevention of seepage through the walls and piping below their foundations.

The opportunities for significantly reducing the use of primary aggregates are limited given the small quantity of materials involved, at least for walls that have a narrow cross-section. It may be possible, however, to build structures intermediate between flood walls and embankments using alternative materials. One design for such a wall using redundant railway sleepers is being developed for a site in Essex. As previously discussed, the replacement of primary aggregates in the concrete elements of such schemes using secondary aggregates should be encouraged.

2.2.3 Bank protection

Protection of the banks of channels, rivers or tidal inlets can be carried out in two main ways. The most direct technique is to strengthen the surface of the bank so that the forces caused by the flows do not damage it. The alternative is to alter the flows so that the currents along the bank are reduced to an acceptable speed. Annual expenditure in the UK on the former class of protection is substantial. Under the auspices of river flood protection alone, the replacement costs for such works are in the region of £20 million a year (see Appendix 1). In addition to this, there are privately funded schemes with a similar purpose. In comparison, schemes to deflect flows away from riverbanks are much less common. The following two sections consider the techniques applied in greater detail.

Revetments

A comprehensive review of the revetments used to protect river and channel banks is provided by Escarameia (1998). A wide variety of techniques have been used for this purpose. Where only a modest amount of protection is needed, then soil reinforcement using various grades of geotextiles is often the most appropriate approach. Where a greater degree of protection is needed, however, more substantial armouring is applied, usually requiring aggregates. Methods used along the banks of channels, rivers and tidal inlets in the UK include:

- asphalt/bitumen-bound sand or crushed stone
- flexible mattresses (eg concrete blocks bound to a synthetic fabric cloth)
- grouted or hand-pitched stone
- concrete blocks (pre-cast and cable-tied or interlocking units)
- rock-filled gabion boxes or mattresses
- rip-rap, ie layers of randomly placed stones, typically up to 250 mm diameter
- fabric bags filled with soil, sand or lean-mix concrete
- encasement of the bank in concrete.

These methods can also be used for the protection of the faces of flood embankments.

Because the main function of such techniques is to resist shear stresses, rather than the direct impact forces that are experienced on coastal slopes, the need for high-specification concrete or stone is reduced. For most of the methods listed there is the potential to reduce the use of primary aggregates. For example, a wide range of secondary aggregates could be used bound in concrete or bitumen, while fill for gabions or under-layers for rip-rap could be provided from recycled aggregates, such as C&D or ceramic waste.

Reducing flow velocities

An alternative to protecting banks to resist the effects of strong currents is to install measures to reduce flow velocities. These structures may be impermeable, and hence be designed to deflect a current, or may be permeable, reducing current speeds by increasing the roughness and dissipating the current energy using frictional effects. The former impermeable type of structure generally requires greater strength to resist the effects of the currents.

Typical flow control measures include:

- groynes, spurs and jetties (ie flow barriers of various lengths and heights built out from the bank)
- guide banks (eg to direct flows and prevent the formation of eddies around bridge piers etc)
- sluices, drop-boards, control gates etc
- drop structures, ie weirs and sills.

Only the last two of these measures are frequently used in flood defence schemes in the UK (see Appendix 1), and these are generally used to control water levels rather than to reduce flows and hence protect riverbanks. Most of these structures could provide opportunities for the use of alternative construction materials, for example replacing rock with recycled aggregates or recovered materials from construction/demolition works such as concrete railway sleepers or blocks.

2.2.4 Bed protection

In some situations, protection of the bed of a channel, river or tidal inlet is required, usually in the vicinity of structures. Typical applications include:

- scour protection at bridge piers and abutments and downstream of weirs and sluices
- covering of pipelines, cables etc where these cross the watercourse
- outfalls, eg at the mouth of an outfall into a watercourse
- intake structures.

The methods used to prevent scour in these situations are similar in many ways to the methods used for bank protection, and can involve flow-calming measures as well as increasing the strength of the bed to resist the strong currents and prevent erosion.

These underwater protection methods can be divided into flexible and rigid categories, the former being able to adjust to modest changes in the underlying bed surface. At the lightest end of the scale come vegetation (eg willow) woven into fascine mattresses and pegged into the bottom, or weighted down with rocks. Applications using primary aggregates include:

- flexible concrete mattresses (eg blocks bound to a synthetic fabric cloth)
- fabric bags filled with sand, lean-mixed concrete etc
- rock-filled gabion boxes or mattresses
- rip-rap, ie layers of randomly placed stones, typically up to 250 mm diameter
- asphalt/bitumen-bound sand or crushed stone.

As with bank protection methods, there is a role for secondary and recycled aggregates, including inert C&D waste to replace the sand, gravel and stones typically used in such schemes.

2.3 SUMMARY OF POTENTIAL USES OF ALTERNATIVE AGGREGATES

The preceding sections have identified a wide variety of coastal and river engineering works that have the potential to use alternatives to primary aggregates. The matching of recycled and secondary aggregates with particular applications, however, requires consideration of the characteristics of those materials, as discussed in the following chapter. The most promising combinations of materials and their applications in coastal and river engineering works are summarised in Table 3.6.

In general, the use of alternatives to primary aggregates in such schemes will require the same permissions or licences normally required, for example under the Food and Environment Protection Act and the Coast Protection Act. There may also be a need to obtain a waste management licence. This issue is further discussed in Section 4.5.

3 Materials availability and suitability

This chapter gives a recent historical account of the availability of the alternative materials under consideration and then an assessment as to their suitability for application in civil coastal and river engineering projects.

3.1 AVAILABILITY OF RECYCLED AGGREGATES

3.1.1 Construction and demolition waste

In 2001 in England and Wales, the construction industry produced an estimated 93.91 mt of construction and demolition (C&D) waste, of which 38.02 mt was recycled as aggregate by crushing and/or screening and 7.05 mt was recycled as soil. Of the remaining 48.84 mt:

- 2.68 mt comprised uncontaminated hard C&D waste and heavily mixed and/or contaminated hard C&D waste with varying potential for recycling as aggregate

- 5.51 mt was mixed construction and demolition excavation waste (CDEW), which was primarily soil but mixed with some hard C&D waste. This had limited scope for recycling as aggregate

- 40.65 mt was wholly or mainly accounted for by waste soil and excavation waste with little or no scope for recycling as aggregate.

The influences of the landfill tax and the aggregates levy can be expected to encourage further recycling of the large amounts of C&D waste that presently are not reused.

Table 3.1 (Environment Agency, 2003) shows that the South East, including London, handles the most C&D waste and recycled and reused the largest. The North West region was the next largest producer followed by the Yorkshire and the Humber region (see source for details of regions). Wales and the North East of England produced the least C&D waste.

Table 3.1 *Destinations of construction and demolition wastes in England and Wales in 2001 (000 t)*

Region	Recycled soil and aggregate	Reused on landfills	Recovered inert at exempt sites	Landfill disposal	Used to backfill quarry voids
East of England	5912	1186	519	475	1294
East Midlands	4859	1048	3129	431	1113
London	4859	218	444	151	379
North East	4247	739	1217	323	937
North West	5352	917	3366	381	1039
South East	5843	1792	2828	779	2202
South West	3579	854	6328	479	1375
Wales	1788	662	1279	352	937
West Midlands	4277	1042	1808	400	1097
Yorkshire and the Humber	4353	950	2764	451	1158
Total	**45 069**	**9408**	**23 682**	**4222**	**11531**

A pilot survey conducted in Scotland estimated (see Table 3.2) that in 1999, of 8.72 mt of C&D waste, 52 per cent went to landfill and the remaining 48 per cent was recycled. Additionally, 60 per cent of soil and rock arisings went to landfill and 40 per cent recycled (DETR, 1998).

Table 3.2   *C&D waste destinations in Scotland 2000*

Material type	Annual arisings (mt)	Landfilled (mt) / *Per cent of arising*	Total recycled (mt) / *Per cent of arising*
C&D waste	2.52	1.26 / 50	1.26 / 50
Soil and rock	4.90	2.94 / 60	1.96 / 40
Estimated total	7.42	4.20 / 57	3.22 / 43

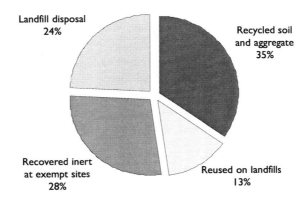

Figure 3.1   *C&D waste destinations, England and Wales, 1998–1999*

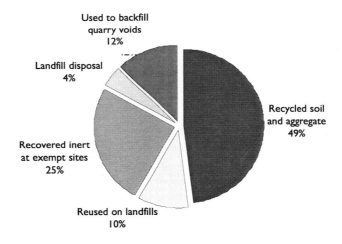

Figure 3.2   *C&D waste destinations, England and Wales, 2001*

3.1.2   Spent railway track ballast (RTB)

Most ballast that is dug out from beneath railway tracks is used as replacement ballast after screening, cleaning and where necessary crushing. Spent ballast (ie that not reused as track ballast) goes for alternative use, generally as graded aggregate after crushing.

Network Rail purchases track ballast from suppliers, then returns it to them for recycling or disposal when the track is upgraded. Only a small portion of the waste material goes to landfill, usually because it is contaminated.

Spent ballast is recycled at 11 centres throughout England and at others in Wales. Working stockpiles exist only at dedicated local sites and only small amounts are available (ODPM, 2001).

3.2 AVAILABILITY OF SECONDARY AGGREGATES

Data gathered in 2001 by the ODPM, and research by others (such as WRAP), has identified locations, volumes of arisings, usage and stockpiles of secondary aggregates in England and Wales. This included materials from existing industrial and construction processes. The scale of reuse/recycling of materials was also recorded where known.

There are no equivalent reported statistics available for Scotland, but some information has been derived from the Scottish Executive Central Research Unit report on recycled aggregates (Winter and Henderson, 2001) based on 1998 estimates and a pilot survey conducted in 2000.

This section reports the existing and future availability for use of the following secondary aggregates:

- china clay sand

- slate aggregate

- colliery spoil

- scrap tyres

- materials arising from capital or maintenance dredging

- glass cullet

- ceramic waste

- spent foundry sand

- blast-furnace slag (BFS)

- basic oxygen furnace (BOF) steel slag

- electric arc furnace (EAF) steel slag

- pulverised fuel ash (PFA)

- power station furnace bottom ash (FBA)

- municipal solid waste incinerator bottom ash (IBA).

Summary tables containing the attributes (as known) for each of these aggregates can be found in Appendix 2. The availability and general suitability of these aggregates for construction purposes is reviewed below. Note that the stockpiles were described as "Usable" or "Relatively hard-to-use", and this book has maintained this terminology.

By-products from mineral extraction operations

All stages of quarrying produce by-products. The materials considered here are:

- china clay waste, which includes coarse sand, waste rock, overburden and a fine micaceous residue left after the extraction of china clay

- slate waste, which consists largely of slate rock left after the quarrying and fashioning of slate roofing tiles and other products

- colliery spoil, the waste from the mining of coal.

Care should be taken in the use of by-products from the extraction of metal ores as these often have high concentrations of metallic compounds that may be hazardous to public health or to plants and animals if used as fill or as cover materials.

3.2.1 China clay sand

China clay extraction occurs predominantly in the south-west of England. Arisings amount to 22.6 mt/a, of which 2.28 mt/a is screened and graded to produce usable aggregate and sand. Coarse china clay sand is suitable for use as fine aggregate, concrete or concrete block manufacture. The remaining sand and gravel is suitable for construction and bulk fill. About 50 per cent of the aggregates used in Cornwall are supplied from this source. There is currently no use for micaceous residue (11 per cent of arisings), which is disposed of by allowing it to settle out in lagoons.

Traditionally, china clay wastes were dumped on pyramid-shaped tips with steep inclines. Often other waste associated with the extraction process, such as machinery and plant parts, are incorporated in them. Latterly, disposal practices have included separate tipping, but these tips may still be comprised of mixed materials associated with extraction, such as sand, gravel, rock and overburden. If these stockpiles are to be utilised then the practicality and economy of separating the fractions may need careful consideration.

Much of the china clay sand could be used profitably if there was sufficient demand for it within its economic area of distribution. Roughly 9 t of waste are generated for every tonne of good china clay (typically 4 t of coarse china clay sand, 4 t of coarse rock material (stent), and 1 t of micaceous residue).

The total quantity stockpiled is estimated at 600 mt, but some of the stockpiles now lie within areas of nature conservation interest. No detailed survey has been undertaken of the extent of the accessible and higher-quality stockpiles that undoubtedly exist – estimates range from 45 mt to 100 mt.

3.2.2 Slate aggregate

Slate waste is by far the largest untapped secondary aggregate resource in the country. Usable stockpiles are estimated as:

- unspecified in the North West (Cumbria)

- 0.5 mt in the South West (Cornwall)

- 277 mt in North Wales (estimated by Arup as being recoverable without undue environmental effect).

Relatively hard-to-use stockpiles are estimated as:

- 5 mt in Cumbria

- 12 mt in Cornwall

- 453 mt in North Wales.

Arisings from the North West, South West and North Wales amount to 6.33 mt/a, of which 0.58 mt is reused as aggregate in low-value applications such as road sub-base, pipe bedding, drain stones and as aggregate for coated macadam road-base and concrete products. It is also used as a low-grade general fill.

Slate wastes are usually loose-tipped on land adjacent to the quarry. Historically, back-filling of old mine workings was also a disposal option (Baldwin *et al*, 1997).

As the material is exempt from the Aggregates Levy, transporting it out of North Wales (the location of about 453 mt of the UK's 456.5 mt total stockpile) is becoming commercially feasible. Alfred McAlpine (the main producer of slate waste in North Wales) anticipates that local road-delivered secondary aggregates will increase by 40–50 per cent over the next few years. Plans exist to invest in railway infrastructure to transport the material to north-west England, the Midlands and possibly the South East (ODPM, 2001).

3.2.3   Colliery spoil

Colliery spoil is the waste from the mining of coal and is deposited in large tips, traditionally in heaps close to the pit heads and processing plants although off-site tipping became more common in later years. Older tips often ignited and the materials in these are generally a mixture of unburnt, partially burnt and burnt spoil. Tips formed more recently comprise predominantly unburnt material and are usually constructed in layers compacted by earthmoving equipment to improve stability and to prevent ignition.

The composition of the spoil varies according to its origin, according to whether it comes from one or several coal seams, derives from the driving of roadways through non-coal-bearing rock, has been processed by a coal washery, or was burnt in the tip. The properties of colliery spoil can vary considerably, therefore, both within the spoil heap and from tip to tip. This should be borne in mind when sampling or chemically analysing colliery spoil.

The rock types most commonly found in colliery spoil are mudstones, siltstones, shales, seat earths, sandstones and, in some areas, limestones. The waste from coal seams contains carbonaceous matter and unseparated coal. The coarser waste from the coal washeries dumped directly from the mine has a wide range of grading, from large lumps of rock to clay fraction size. Some breakdown of the materials in spoil is likely. When first deposited, the unburnt spoil is usually grey or black and has a significant clay mineral content. When burnt it usually becomes reddish (though lack of oxygen during cooling may produce a blackish colour) and is mechanically stronger. The fine "tailings" from coal recovery plants are often deposited in lagoons on the tip. They tend to contain more sulphate minerals and carbonaceous matter (BS 6543:1985).

Usable stockpiles amount to about 20 mt and are only found at working mines. However, planning restrictions make significant usage of even the most available of stockpiles unlikely, unless they coincide directly with development.

Exemption from the Aggregates Levy has improved the commercial feasibility of using colliery spoil as aggregate. It is the view of the industry that any moves towards performance specifications rather than "recipe" specifications in construction would also assist (ODPM, 2001).

3.2.4   Scrap tyres

Scrap tyre arisings amount to some 400 kt/a in the UK. Around 90 kt is processed into rubber crumb before being used as aggregate substitute in the surfacing of sports and play areas and in road surfacing. Some whole tyres are used in landfill engineering and a significant number are used to hold down plastic sheeting on silage clamps on farms. Of the 170 kt put to non-aggregate use, 20 per cent is used as fuel in cement kilns, 10.5 per cent is recycled as retreads, 8 per cent incinerated with energy recovery, 2 per cent in carpet underlay, 2 per cent exported and 1 per cent used in pyrolysis.

Quantities are increasing by 2 per cent a year and the recent ban on landfilling tyres will increase those available for recovery. During 2002, as part of the Environment Agency's waste tyres programme, AEA Technology was commissioned to carry out a tyre stockpile survey. This survey indicated there are some 38 stockpiles of 1000 or more tyres in England and Wales, with a total of nearly 14 million tyres (AEA Technology, 2002).

3.2.5 Maintenance dredged materials

A dredged material is that which is "recovered by the dredging process from any surface water body and which the operator needs to dispose of" (Waste Management Licensing Regulations 1994). The source of the material may be from any watercourse, estuary, port, canal or lake.

Dredged material disposed of annually amounts to 40 mt, mainly to licensed sites in estuaries and offshore. Typically, about 80 per cent of the material arises from maintenance dredging, and is mostly muddy sand.

3.2.6 Glass cullet

Usable stockpiles of glass cullet are very small and organisations have to build up stocks before they can market recycled products. There are no long-term stockpiles because virtually no glass is segregated and landfilled. The 1.3 mt a year placed in landfill as mixed waste cannot be recovered.

Recycled waste glass is segregated at source, separated or collected from bottle banks. An estimated 2.2 mt of post-consumer container glass (ie not including waste glass from end-of-life vehicles, C&D waste and flat-packed glass) passes through the municipal solid waste stream per year. Fluctuations in arisings are dependent upon product packaging trends, consumer demand and on public participation in glass recycling schemes.

Generally it is deemed more appropriate to recycle container glass back into new containers rather than use it for any other purpose, and subsequently only a few thousand tonnes are used in glass-fibre manufacture and for other uses such as water filtration. Established aggregate uses include sharp sand for bedding beneath block paving and pipework, abrasive uses (shot-blasting), non-structural pre-cast concrete, "glasphalt" and landscaping products.

3.2.7 Fired ceramic waste

Although potentially useful as a fill material or even for beach recharge, utilisation of fired ceramic waste is already high and there is little scope for increase due to efficiency gains and increasing utilisation within the ceramics sector itself (ODPM, 2001).

Fired ceramic waste arises from the brick, tableware and sanitary industries nationwide, with local concentrations in the West Midlands and east of England. Arisings amount to only 0.1 mt/a, of which 90–100 per cent is reused for bulk fill, crushed for aggregate and incorporated into brick-pit and landfill haul roads.

3.2.8 Spent foundry sand

The uses for spent foundry sand are relatively established in block manufacture and ready-mix concrete. A small amount is used in asphalt manufacture (ODPM, 2001).

Spent foundry sand arises from the castings industry located mainly in the Midlands and the north of England. Annual arisings amount to about 0.9 mt, of which only about 10–20 per cent is reused. Further decline in arisings is likely due to the reduction in the castings industry. There are no reliable estimates of sizeable existing usable stockpiles, although anecdotal evidence suggests that there are some substantial dedicated landfills of foundry sand at major castings works possibly amounting to several million tonnes.

3.2.9　Metallurgical slags

The main ferrous slags in the UK are blast-furnace slags (BFS) from the production of iron, and steel slag from the production of steel. Blast-furnace slags are relatively homogeneous and the molten slag may be processed by:

- cooling in air to give a rock-like material
- cooling with a limited supply of water (foaming or pelletisation) to produce lightweight granular material
- cooling with an excess of water (granulation) to produce a glassy granulated material.

With the exception of material in poorly characterised historic waste tips, all BFS is now reused in some form. It is used as a high-quality material in the cement industry, in glass-making, concrete block manufacture and for roadstone applications. These are all premium products commanding an appropriate price.

The steel slags produced today are denser and more variable in composition than BFS. They originate mainly from either the basic oxygen steel-making process (BOF or BOS) or from the electric arc process (EAF) and produce different slags. Older slags may be from the open hearth process. Steel slags have properties quite distinct from those of blast-furnace slag and may contain residual iron, free lime (CaO) or free magnesia (MgO); the last two may make them unstable. To avoid instability problems, steel slags are allowed to weather to hydrate the residues and stabilise the material before use. As a precaution, steel slags should not be used where failures can be caused by any remaining instability, eg as fill under buildings or as aggregate in concrete.

In assessing the use of non–ferrous slags, consideration should be given to possible toxicity from heavy metals. Tin slag can be found on a beach in Liverpool Bay predominantly made up of C&D waste. This beach is deemed to be contaminated by the tin slag (Masters, 2002) (see Case Study A3.7). Pollution studies have been recommended to establish the effects and extent of any impacts, and to help develop any necessary remedial measures. In addition, any tin, copper and lead-zinc slags are produced in relatively small quantities. In view of the above comments, these materials are not considered further in this study.

3.2.10　Blast-furnace slag (BFS)

BFS waste arisings amount to around 3 mt/a from Yorkshire and the Humber and South Wales. Between 60 and 70 per cent of quenched slag is ground up and used as ground granulated blast-furnace slag (GGBS), a cement replacement (ie not an aggregate). Air-cooled slag can be used as roadstone aggregate and this share is estimated at 30–40 per cent (ie the difference).

A decline in slag arisings and strong growth in the GGBS market has reduced the amount of BFS available for aggregate use. During the 1990s, BOF steel production was relatively stable (~13 mt/a) but by 2001 it had declined to 10.4 mt/a and further decline is likely due to steel industry trends (ODPM, 2001).

3.2.11 Basic oxygen furnace steel slag (BOF)

BOF steel slag arises from integrated steel works (those that contain both blast furnaces for iron processing and basic oxygen furnaces for steel-making) in Yorkshire, Humberside and South Wales.

Arisings amount to about 1 mt/a, of which 98 per cent is reused as aggregate mostly as roadstone in England. The remaining 2 per cent is used in agriculture. There are no reliable estimates of stockpiles and it is quite likely that none exist, as BOF steel slag is a high-quality material with a well-established use for which demand outstrips supply. Thus the quantity used for aggregate is governed by arisings (ODPM, 2001).

3.2.12 Electric arc furnace steel slag (EAF)

EAF slag arises from electric arc furnace steel plants in Yorkshire and the Humber and the South East. Arisings amount to about 0.28 mt/a, all of which is reused as aggregate. It is a high-quality material used in asphalt mixes for roads and surface dressing where skid resistance is important. There has been a gradual decline in EAF slag arisings reflecting the decline in steel production in the UK, a trend that is likely to continue. There are not thought to be any long-term stockpiles (ODPM, 2001).

3.2.13 Pulverised fuel ash (PFA)

Pulverised fuel ash is the fine powder removed from the exhaust gases of power stations that burn pulverised coal. PFA accounts for 75–80 per cent of all ash from coal-fired power stations, with the remainder being heavier, agglomerated particles known as furnace bottom ash (FBA).

PFA varies in colour from light to dark grey, depending on the unburnt carbon content. Much of the ash is deposited in stockpiles or lagoons. Alternatively, the ash may be held in hoppers at the power station, available either as free-flowing dry powder or conditioned with water to a semi-damp mass. When deposited, the latter may gradually harden to produce material like soft sandstone (BS 6543:1985).

Arisings amount to around 4.87 mt/a, of which about 1.66 mt is used as aggregate and 0.83 mt has non-aggregate uses. Usable stockpiles located at coal-fired power stations are estimated at approximately 55 mt. In addition, many more millions of tonnes of PFA could be accessed by stripping topsoil from covered stockpiles, if this proved to be economically worthwhile.

3.2.14 Furnace bottom ash and clinker (FBA)

Furnace bottom ash is the coarser agglomerated material recovered from the bottoms of the combustion chambers of power station boilers fired with pulverised fuel.

It ranges from a highly vitrified, glossy and heavy material to a lightweight, open-textured and more friable type. The precise nature of the material will depend on the boiler plant and coal type. It is sometimes found mixed with PFA in stockpiles.

The production of clinker from power station boilers using chain grate stokers has almost ceased, but some may still be obtained from other sources (BS 6543:1985)

At present, most FBA (like PFA) is sold, mainly for the manufacture of building blocks and for bulk fill applications (generally in road construction) and no substantial usable stockpiles of this material exist.

3.2.15 Municipal solid waste incinerator bottom ash (IBA)

The direct incineration of domestic and trade refuse leaves a clinker material, containing iron and other metals, glass and cinders together with smaller amounts of unburnt paper, rags and vegetable matter.

Municipal solid waste incinerators exist in the North West and North East, Yorkshire and the Humber, East and West Midlands, east of England and London. Municipal waste is increasing and the EU Landfill Directive requires much more to be diverted from landfill over the next 20 years. The current capacity for municipal waste incineration is 2.7 mt/a, less than 10 per cent of the waste produced. The Environment Agency estimates that the amount of waste incinerated or recovered by other means may need to double to reach the target of 10 mt/a by 2010. This could result in IBA arisings of about 2.5 mt/a by 2010.

Working stockpiles are small and short-term. Stockpiling of IBA requires a waste management licence. However, driven by the increasing cost of disposal and exemption from the Aggregates Levy, its use as an aggregate is likely to increase so long as it can be shown to be safe to use and that any regulatory constraints are not too onerous.

The main uses of IBA are for road surfacing (in asphalt) and concrete block-making as well as bulk fill, road-base material and for daily cover at landfill sites (ODPM, 2001).

3.3 SUMMARY OF AVAILABILITY OF SECONDARY AGGREGATES

In order to identify those materials most likely to be shortlisted for further consideration it is useful to compare their attributes as discussed in Section 3.2. Table 3.3 below illustrates the tonnages of annual arisings available for use for each material.

As previously described, most of the candidate materials already have substantial well-defined uses. Materials are stockpiled if they do not have a market use, where arisings outstrip demand, or when they are of too low a quality.

These stockpiles are the reserves available for coastal and river engineering applications. Table 3.3 lists the tonnages of stockpiles potentially available for utilisation in England, Wales and Scotland. It is apparent from the figures that the volumes of secondary aggregate available in Scotland are far lower (thousands of tonnes) than those in England and Wales (millions of tonnes).

Table 3.4 below summarises for each material what percentage of their annual arisings is already reused (England and Wales only).

Those highly utilised, and therefore scarce, materials shown in Table 3.4, such as BFS, PFA and BOF steel slag, command a market value unlikely to be affordable for many applications in coastal and river engineering schemes. Those further down the table, such as tyres, china clay sand, colliery spoil and slate aggregate, are more affordable because of their under-utilisation and are a cheaper option for high-volume schemes.

Table 3.3   Tonnage of materials in England, Wales (2001) and Scotland (1998)

Resource type	Material	Annual arisings England and Wales	Annual arisings Scotland	Potential aggregate portion England and Wales	Potential aggregate portion Scotland	Actual aggregate use England and Wales	Actual aggregate use Scotland	Non-aggregate use England and Wales	Non-aggregate use Scotland	Stockpiles England and Wales	Stockpiles Scotland
Metallurgical slags	Blast-furnace slag	3.0 mt	0	3.0 mt	Unknown	0.9–1.2 mt	90 kt	1.8–2.1 mt	Unknown	No reliable estimates	0
	BOF steel slag	1.0 mt	Unknown	1.0 mt	Unknown	0.98 mt	Unknown	0.02 mt	Unknown	No reliable estimates	0
	EAF steel slag	0.28 mt	Unknown	0.28 mt	Unknown	0.28 mt	Unknown	0	Unknown	No reliable estimates	Unknown
Mine and quarry	China clay	22.6 mt	0	20.01 mt	0	2.28 mt	0	0	0	45100 mt	0
	Slate	6.33 mt	Unknown	6.33 mt	Unknown	0.58 mt	Unknown	0	Unknown	456.5 mt	Unknown
	Colliery spoil	7.52 mt	150 kt	7.52 mt	150 kt	0.81 mt	65 kt	0	0	10–20 mt	Unknown
Other	Pulverised fuel ash	4.87 mt	780	4.87 mt	780 kt	1.66 mt	228 kt	0.83 mt	Unknown	55 mt	Unknown
	Furnace bottom ash and clinker	0.98 mt	44 kt	0.98 mt	44 kt	0.97 mt	40 kt	0	0	No reliable estimates	Unknown
	Incinerated refuse	0.62 mt	Unknown	0.62 mt	Unknown	0.38 mt	Unknown	0	Unknown	No reliable estimates	Unknown
	Spent railway track ballast	1.3 mt	102 kt	1.3 mt	102 kt	1.24 mt	77 kt	0	Unknown	No reliable estimates	Unknown
	Spent foundry sand	0.9 mt	Unknown	0.9 mt	Unknown	0.09–0.18 mt	Unknown	0	Unknown	No reliable estimates	Unknown
	Glass waste	2.2 mt	Unknown	2.2 mt	Unknown	85 kt	Unknown	0.65 mt	Unknown	20–30 kt	Unknown
	Fired ceramic waste	100 kt	Unknown	100 kt	Unknown	90–100 kt	Unknown	0	Unknown	Working only	Unknown
	Scrap tyres	400 kt	Unknown	400 kt	Unknown	90 kt	Unknown	170 kt	Unknown	~14 million tyres	Unknown

Notes

Figures from: Construction and demolition waste survey: England and Wales 1999/2000 (ODPM, 2000), Survey of arisings and use of secondary materials as aggregates in England and Wales 2001 (ODPM, 2001), Recycled aggregates in Scotland (Winter and Henderson, 2001)

Table 3.4   *Annual production reused*

Materials	Proportion of annual production reused (2001)
Blast-furnace slag	100 per cent
Electric arc furnace steel slag	100 per cent
Fired ceramic waste	90–100 per cent
Power station furnace bottom ash	99.3 per cent
Basic oxygen furnace slag	98 per cent
Used tyres	65 per cent
Municipal solid waste incinerator bottom ash	61.3 per cent
Pulverised fuel ash	51 per cent
Glass cullet	33 per cent
Spent foundry sand	10–20 per cent
Colliery spoil	11 per cent
China clay sand	11 per cent
Slate aggregate	9 per cent

3.4   STANDARDS AND SPECIFICATIONS

Before starting any appraisal of the possible uses of a material type, there are measures of particular characteristics that need to be gathered to assess its likely structural and functional performance. These include a combination of the following, depending on the material type and its intended purpose.

Box 3.1   *Measures of material characteristics*

Porosity	Effective angle of friction and effective cohesion
Permeability	Frost heave
Solubility	Leachants
Particle size	Water absorption
Strength	Nitrates content
Hardness	Redox potential
Durability	Chloride ion content
Uniformity	Total sulphate content
Grading	Acid-soluble sulphate content
Density	Sulphide and hydrogen sulphide
Colour	Moisture condition value (MCV)
Flakiness	Grain compression/tension
Plasticity	Undrained shear strength parameters
Elasticity	Susceptibility to chemical or biological attack
pH value	Optimum moisture content and maximum dry density

This is not a comprehensive list, but it indicates the level of scientific investigation that may be required to establish material characteristics for construction purposes. Note that these characteristics may need to be established for primary aggregates just as much as for alternative materials; there is no justification for assuming that alternative materials would be inferior without specific evaluation. The specification of materials with certain properties and characteristics is necessary to ensure that design, maintenance and repair of a structure conform to tried and tested standards. These standards ensure that structures perform the desired function(s) and safeguard human and environmental health.

Specific material characteristics, physical properties and performance, and their specifications (where available) can be found in CIRIA publication C513 *The reclaimed and recycled construction materials handbook* (Coventry, Wolveridge and Hillier, 1999). However, there are few case studies illustrated in that publication that relate to coastal or river engineering applications.

Many engineers are, understandably, reluctant to consider the use of recycled or secondary aggregates, or even to allow their use, unless specifications are available. Among the reasons for this are concerns about durability of materials, a perception that there may be an increased risk of adverse environmental impact and an unwillingness to jeopardise the consultants' professional indemnity insurance to clients. This has fostered a highly conservative attitude to the choice of materials in the civil engineering industry in the UK.

Clearly defined "fit for purpose" specifications can help alter the perception that the use of recycled and secondary aggregates in coastal and river engineering is another form of waste disposal. They may reassure those concerned that material characteristics and behaviour have been investigated and assessed, under appropriate conditions, to prove that the materials are a valuable and suitable commodity in the construction market, posing no threat to the environment.

Under such circumstances, there is clearly a need for major public-sector bodies such as the Environment Agency to take a lead in promoting the use of recycled materials. This may require such bodies to accept some, if not all, of the risks associated with the use of secondary or recycled materials.

The new European Standards that came into force on 1 January 2004 will replace British Standards for aggregates (eg BS 6543:1985 *Guide to the use of industrial by-products and waste materials in building and civil engineering*), which will be withdrawn in June 2004.

The new Standards will have a positive effect on the use of secondary and recycled aggregates because they are included in the scope of the new standards on an equal basis with primary aggregates. Although the aggregates remain the same, their terminology, product descriptions, standard sieve sizes, grading presentations and test methods have changed. A list of the new Standards and national guidance documents is shown at the end of the References section and are available from the BSI. These include guidance notes ("PD" documents) for the new EN standards.

Gaps still remain in the relevant clauses, however, and use in the near future will still depend upon existing and new UK guidance.

The Building Research Establishment is compiling a report on "fit for purpose" specifications, which will include marine and river construction works and is aimed at encouraging the use of secondary and recycled materials. In particular it will draw on experience from the USA and Europe in applications for roller-compacted concrete (RCC), concrete-bound materials (CBM) and hydraulically bound materials (HBM). The guidance is to include the principles by which soil and aggregates (including recycled and secondary aggregates) that are well established in road construction can be used in other applications such as shore and slope stabilisation and erosion control.

The Environment Agency and WRAP are drafting a new quality protocol for producers. This will enable the producers to declare that a product conforms to specifications and is of the required quality. Guidance on the new European Standards has been widely disseminated – by the Quarry Products Association and the British Standards Institution (BSI), among others – and most producers are already implementing them.

Existing British Standards of direct relevance include BS 6349. Parts 1 to 7 cover the construction and design criteria for maritime structures such as:

- jetties
- shiplifts
- quay walls
- dock and lock gates
- dolphins
- fendering and mooring systems
- dry docks

- dredging and land reclamation
- locks
- inshore moorings
- slipways
- floating structures
- shipbuilding berths
- breakwaters.

However, these standards refer to very few secondary or recycled materials and there do not appear to be equivalent British Standards for river structures.

The absence of specifications or standards for river structures such as flood embankments is of some concern. A flood defence is constructed specifically to prevent the passage of floodwater. Not all embankments are constructed for this purpose. For example, road and railway embankments are not normally designed to prevent the passage of floodwater and can suffer serious damage if subjected to high floodwater levels. Railway embankments are particularly susceptible to damage from high floodwater levels. Many were constructed during the Victorian era and were built from a wide range of relatively poor-quality and poorly compacted materials. Seepage through or inundation of railway embankments can lead to erosion or collapse.

There is a range of good practice guidance available (see Box 3.2) for flood defence embankments, but apparently no national standard for acceptable materials. The provision of standard guidance is to be recommended in a good practice review report (EA, in prep). Until this guidance is available (in 2004), recommendations as to the suitability of materials can only be based upon basic structural engineering principles and known performance.

Box 3.2   *Good practice guidance available for flood defence embankments*

CIRIA C592 *Infrastructure embankments – condition appraisal and remedial treatment*, 2nd edn (Perry, Pedley and Reid, 2003)

Dikes and revetments. Design, maintenance and safety assessment (Pilarczyk, 1988)

An engineering guide to the safety of embankment dams in the United Kingdom, 2nd edn (Johnston et al, 1999)

Investigating embankment dams – a guide to the identification and repair of defects (Charles et al, 1996)

R&D Publication 11, v 1.0 *Waterway bank protection: a guide to erosion assessment and management* (Environment Agency, 1999)

CIRIA Book 9 *Protection of river and canal banks* (Hemphill and Bramley, 1989)

River and channel revetments – a design manual (Escarameia, 1998)

The earth manual, 3rd edn (United States Bureau of Reclamation, 1998)

Embankments on soft clays (Leroueil, Magnan and Tavenas, 1990)

In the absence of sufficient information, consideration of suitability can only be based upon known properties of the materials and their performance in similar applications.

There are some general specifications with respect to earthworks from Viridis (Reid, 2003) that may for instance apply to flood embankments. Viridis has also developed a set of new protocols in the form of a series of linked flow-charts that take the user through the questions that must be addressed before a material can be used in any application. Where a material cannot be used, measures that need to be taken to enable its use are given or alternative uses suggested.

Equivalent guidance of this nature is lacking for coastal and river engineers and is an area where much more work needs to be done, as existing protocols are not necessarily accepted universally throughout the industry. Materials testing, modelling and use in pilot projects need to be conducted in coastal and river engineering schemes to ascertain suitability. Some studies using tyre bales are being conducted in the UK to this end (see Section 3.5.6).

Parts of BS 6349 demonstrate how specifications can promote the use of alternative materials by stating performance benefits and clear caveats to their selection. Section 7 (Materials) of BS 6349-1:2000 *Maritime structures. General criteria* refers to the beneficial use of ground granulated blast-furnace slag and pulverised fuel ash in "marine" reinforced concrete where they help to prevent chloride- and sulphate-induced corrosion, and lists the relevant British Standards for each cement mixture. Paragraph 58.4.2 also states that aggregates (for concrete) should conform to BS 882:1992 or BS 3797:1990 for lightweight aggregates. It also states that if recycled materials are to be considered then studies of petrography, appropriate physical and chemical tests, and trial mixes should be made.

The quality of aggregates has a lesser impact on the strength of concrete than is often supposed (other than effects of water demand, which might affect the water-cement ratio), but it does affect resistance to abrasion, freeze-thaw and surface weathering. If severe abrasion of the concrete by pebbles or sand is expected, the coarse aggregate should be at least as hard as the material causing the abrasion and the fine aggregate content of the mix should be kept as low as is compatible with appropriate mix design.

Section 7 (clause 65) of BS 6349 also refers to "bituminous" materials for use in marine works, such as:

- pavements
- coatings
- waterproofing
- sealing compounds
- coast and bank protection.

Bitumen is employed in various ways for coast and bank protection work. Common mixtures include:

- asphaltic concrete
- sand mastic or asphalt mastic
- dense stone asphalt
- open stone apshalt
- lean sand asphalt.

Table 3.5 illustrates the extent to which bituminous mixes are used for different functions on different structures.

Again there are is no specific reference to secondary or recycled materials, but these can be used so long as they conform to the specified requirements (which often vary within wide limits).

Types of bituminous material	Dykes and closure dams	Dune protection and seawalls	Protection of seabed	Groynes	Breakwaters	Sills in closure gaps
Asphaltic concrete	Revetment above high water	Revetment above high water level	Aprons placed in the dry	Special cases (capping)	–	–
Sand mastic grouting	Grouting of stone revetment	Grouting of stone revetment	–	Revetment and capping Grouting up stone	Moderately attacked revetment only	Heavily attacked sill
Sand mastic carpet (placed *in situ*)	Toe protection	Toe protection	Plain or stone weighted or roughened	Toe protection	Toe protection	–
Prefabricated mattresses	Toe protection revetment	Toe protection revetment	Special cases	Toe protection	Toe protection	–
Open stone asphalt	Revetment	Revetment either direct or by grouting of heavy rubble	Only prefabricated mattresses	Revetment either direct or by grouting of heavy rubble	Grouting of heavy rubble	–
Dense stone asphalt	–	Revetment	–	Revetment	Revetment	–
Lean sand asphalt	Core, fill, filter layer	Core, fill, filter layer	–	Core	Core	Core

3.5 SUITABILITY OF CONSTRUCTION AND DEMOLITION WASTES AND SECONDARY AGGREGATES

3.5.1 Construction and demolition waste

The Environment Agency has dramatically increased the level of recycled aggregates used in flood defence works over recent years. The Agency used more than 800 000 m³ of construction materials in 1998/9, of which 38 per cent was secondary or recycled construction aggregates (Environment Agency, 2001). For example, 500 t of revetment stone from the Albert Dock flood defence scheme in Hull was reused to protect the Humber bank at Easington. Some 300 t of recycled coping stones were used as sea defence to protect cliffs at Barmston Sea End, East Yorkshire, from erosion, and brick rubble was used in place of new crushed stone on the Beck River works project in the North East.

On the Norfolk coast concrete blocks from demolition works have been used in crib-work for cliff toe protection (see Case Study A3.5). Other waste material forms could potentially fulfil this application, including concrete railway sleepers, old stone block-work, prefabricated concrete sections and kerbstones.

Recycled scalping from haul roads in the works site were used as a free-draining sub-base foundation for flood embankments and as free-draining fill in railway embankment stabilisation works on the Melton Mowbray Flood Alleviation Scheme (see Case Study A3.12).

Construction and demolition (C&D) waste may be considered for a coastal or river engineering application if there are demolition or construction projects locally or within the region that will give rise to enough suitable material. An opportunity to reuse inert material that would otherwise be disposed of or used for lower-value applications should not be missed.

Gabions are often realistic substitutes in concrete applications (see Primrose Wharf, Case Study A3.13). Eliminating problems of availability would go a long way to increasing C&D waste utilisation in this form.

However, there are problems to be addressed when considering utilising C&D wastes for some applications such as beach recharge or gabion and embankment fill, due mainly to the difficulty of maintaining consistency of quality and supply when sourced from different sites.

The Hayling Island Oysterbed case history (Case Study A3.6) illustrates why careful consideration should be given to screening, grading and crushing of the material and as to whether the resulting consistency and quality would fulfil the material specification for the intended application.

It is conceivable that C&D waste could become a reliable source of raw material for small-scale local remedial works if a programme of continuing maintenance of sea and flood defences was under way. A steady demand for a product will make its use more viable. The consistency and quality problems associated with multi-sourcing would of course have to be addressed and overcome. However this is less of an issue now than it used to be with the advent of improved screening and grading capabilities of modern crushing and recycling plant, and the uptake of quality control systems and protocols by the recycling industry (eg BRE, 2000a).

An important consideration is the opportunity for the recycling and reuse of C&D waste from coastal and river structures themselves. In some cases this may prove to be the most cost-effective and environmentally sensible solution (see Minehead, Brighton Marina and Beesands case studies in A3.3, A3.4 and A3.5 respectively). To date there have been few opportunities for recycling or reuse, however, as waste is very rarely, if ever produced. Instead, original structures tend to be incorporated as they exist within new schemes. This reduces the scale of the new works and saves energy that would otherwise have been used in the disposal of the original structure.

3.5.2 Spent railway track ballast

No examples of the use of spent railway track ballast could be found but there is no apparent reason why clean ballast sufficiently graded, sorted and/or crushed if required could not be used. Contaminated ballast would require cleaning and testing to make sure it was fit for purpose.

3.5.3 Slate aggregate

Slate aggregate is inert and, although the tips are usually relatively clean, the material covers a wide range of particle sizes.

Slate has historically been used in coastal structures in Scotland. A good example is the harbour seawall at Easedale (see Case Study A3.18). Once crushed and screened, it also becomes a potential source of shingle-sized material suitable for beach recharge, particularly in those areas of western Britain with similar natural beach material. However, there is little local demand in these areas. The use of slate aggregate in areas of higher demand, such as in the south and east, depends largely on the costs of transportation by rail or sea, although factors such as appearance and durability also affect the choice of material.

Arisings (94 per cent of "good" slate production) appear relatively stable. The English slate industry is entirely quarry-based and small in comparison with that of Wales.

About 85 per cent of English waste arisings are in Cornwall. All major quarries in North Wales have planning consents for more than 20 years' extraction and scope for many more years of operation. Welsh arisings are expected to continue at more than 4 mt/a.

Utilisation of the large reserves in North Wales has been relatively low because of the low quality and value of slate waste, and the reserves' distance from main aggregate markets. Exemption from the Aggregates Levy and potential investment in railway infrastructure means that its use as aggregate is likely to increase. It is anticipated that it could take several years before significant increases are observed, however.

Until the transport infrastructure is in place to "export" slate waste economically to other parts of the country there is little realistic chance of it being used in coastal or river projects other than in North Wales. Furthermore, a transport economic appraisal undertaken for the National Assembly for Wales (Ove Arup, 2001) indicated that without significant subsidy of new infrastructure and railway freight operating costs, the current level of aggregates tax (at £1.60/t) was insufficient to encourage uptake of slate waste in England. It was suggested that a £2–2.50 increase would be required to make bulk transportation an attractive proposal.

The study indicated that £15–21 million would be required for mainline railway links and infrastructure improvements to access the stockpiles at Blaenau Ffestiniog and Penrhyn.

3.5.4 Colliery spoil

Colliery spoil has been used for many years as construction material for impounding dams (for the storage of fly ash) at coal-fired power stations and in lake and reservoir dams.

There are strong similarities between the mechanical properties of colliery spoil and inorganic clays of medium plasticity. Colliery spoil has been successfully used to construct earth reinforced structures using grid reinforcements (Jewell and Jones, 1981). Colliery spoil can be very susceptible to moisture, so proper drainage and shaping of the structure is essential. The material can be satisfactorily compacted at a moisture content close to the optimum, but it may be slightly more difficult to place than other more conventional fill materials (Jones, 1985).

Of the 7.52 mt of colliery spoil arisings from the deep coal mines of the North East, Yorkshire and Humberside, West and East Midlands and South Wales, only 0.81 mt (11 per cent) is reused as aggregate. Most is used as bulk fill, notably for constructing lagoons at collieries themselves and in site restoration. It has also been used in landfill engineering, flood defence works and road construction. At the peak of motorway construction in the 1970s, some 8 mt/a were being used as bulk fill.

The placement of colliery spoil on the beaches (eg the County Durham coast, see Case Study A3.2) was not intended as a beach recharge but provided some information on the possible advantages and disadvantages using this material for such schemes. Since the mines in Durham and Fife have closed, the loss of the colliery spoil from the beaches has been very rapid.

The main problem with the use of colliery spoil is its variability. For this reason, it has until now been suitable only for use as a fill material or other applications where variable quality is acceptable. Higher-grade utilisation demands better uniformity throughout and this upgrading requires costly washing, screening and blending processes.

There is a good track record of colliery spoil being extensively tested and used as embankment fill for flood defence schemes (see the Selby-Wistow-Cawood Barrier bank case study in Case Study A3.16) and also a successful example of shingle displacement fill for a beach in Kent (see Betteshanger Colliery case study in Case Study A3.1).

3.5.5   China clay sand

China clay waste is a material that shows potential for realistic utilisation in the near future. China clay sand is available for use in the construction industry in three forms:

- dump sand – unprocessed material straight from the pit or tip
- screened sand – dump sand, dry-screened to remove oversize material
- processed sand – sand from the pit or dump that is processed through a washing and grading plant.

China clay sand is a potential alternative resource for beach recharge material and has been used for beach reclamation at Saltash (on the Tamar) and at Crinis Beach in St Austell Bay. It has also been used extensively for decades as the construction material for tailings dams containing the china clay industry's own settling lagoons and for general granular bulk fill in reservoir dam embankments in Cornwall such as at Colliford Lake on Bodmin Moor (see Case Study A3.17).

The sand is similar to a Zone 1 (category C) concreting sand (BS 882:1992) with a median grain size of 1–3 mm. It also contains very small amounts of clay and other fines (material < 75 microns). Slopes of beaches of china clay sand in Cornwall can be as steep as 1:7, similar to the profile of shingle beaches, reflecting both the grain size and the narrow grading of the surface sediments. Used elsewhere, however, and mixed with other sediments, less steep gradients are likely to form.

Technically, processed china clay sand is able to meet the required specifications for building and concreting sands and the unprocessed sand is an acceptable fill material.

Cost of transport has been the main constraint on the utilisation of china clay waste beyond its economic area. However, with existing usable stockpiles of about 45–100 mt in Cornwall, exemption from the Aggregates Levy and investment in the Port of Par development, it is becoming an increasingly competitive source of sand and aggregate. The feasibility of moving substantial quantities of material by rail and sea from Cornwall to engineering projects in the South East, the South West and perhaps farther afield in England will be increasingly plausible after completion of the new development. By 2004/5 sales are expected to increase by about 80 per cent, with 30 per cent of total sales being exported by sea (ODPM, 2001).

3.5.6   Used tyres

Use of tyres in coastal and river engineering is far from a new idea. Whole tyres have been successfully used to construct artificial reefs, revetments and breakwaters in low-energy environments in numerous countries, including Australia, the United States, Israel and the UK (see Appendix 3)

Stockpiles of used tyres may not equate in tonnages to aggregates but they are large in number and are becoming an increasing disposal problem and environmental hazard when left exposed in large dumps.

Potential for growth in aggregate substitution exists, but total amounts are likely to remain small. Interest in recycling has been stimulated by the Landfill Directive (which called for the ban on landfilling of whole tyres by July 2003 then of shredded tyres by July 2006). Tonnages used in cement kilns and in pyrolysis are predicted to increase, raising the level of competition with tyre use as aggregates.

Pilot projects into the use of tyre bales in coastal and river engineering are in progress (<http://www.tyresinwater.net>). Several pilot schemes are being installed, studied and sampled to ensure that there are no deleterious effects on the environment from tyres and that the designs are structurally sound. Scientific evidence to date suggests that the risks of adverse impact on the environment with respect to leachates are negligible. If these pilots are deemed a success then potentially significant tonnages could be used for these purposes, due in part to innovative tyre baling techniques and engineering design.

Initial signs indicate that this may be an extremely useful utilisation of tyres, which in structures such as embankments on soft ground may even be a preferable option to conventional designs (see Case Study A3.10).

There are also other methods of utilisation, for example, weaving used tyre treads into versatile matting systems where the tyre walls are removed to leave the tread of the tyre with the metal reinforcing still encapsulated within the rubber. These loops of tread are then woven together (without cutting) into various designs of rubber mats. These mats may be used for applications such as:

- bed protection/scour prevention
- groyne construction
- protective screens for cables and pipes
- artificial reef construction
- large-dimension fender for the protection of a quay wall
- construction of a tidal deflector in a yacht harbour.

3.5.7   Maintenance dredged material

Although about 80 per cent of maintenance dredged material is muddy sand, it is still regarded as a potential resource by FEPA regulators for recharging or recreating intertidal habitats. This practice at present is limited to small-scale trials due to the uncertainty of ecological impacts (see Horsey Island, Titchmarsh Marina and Suffolk Yacht Harbour case studies in Appendix 3).

Rejuvenation of these soft mudflat and saltmarsh coastlines does have legitimate engineering benefits. By efficiently dissipating wave energy, these areas act as buffer zones in front of conventional sea defences. However, relative sea level rise has contributed to the erosion of these areas, especially in south-east England, where salt-marshes are dwindling at a worrying rate. Their disappearance results in direct and increased wave attack on structures behind them often undermining and destabilising defences. This can also result in increased costs for building, repair and maintenance.

This use of dredged material whilst not a direct substitute in terms of its application for primary aggregate, does offset the cost of maintenance and replacement of seawalls and similar structures, and is being increasingly seen as cost-effective way of maintaining sea defences.

Each disposal case is assessed individually by Defra, as each has its own legal, economic, environmental and logistical limitations.

Various CEFAS reports demonstrate further the beneficial uses of maintenance dredged material (see Appendix 3 and <http://www.cefas.co.uk/decode/use.htm>).

Examples of the use of maintenance dredged materials in river engineering, although practised widely, do not seem to be well documented. Accounts of large schemes such as those on the upper Mississippi River, where extensive use of river-dredged (backwater clearance) material has been used to re-create islands and habitat restoration, are available (see <http://www.tellusnews.com/ahr/report_chapter5.html>).

The majority of coastal material dredged is cohesive and unsuitable for beach nourishment or fill applications, although on some high-energy, sandy coastlines it may be relatively clean sand of potential value for beach recharge.

Capital dredging works, ie widening or deepening navigation or berthing areas to a greater depth than previously dredged, can provide more suitable sediments for beach nourishment purposes (see Poole beaches case example in Case Study A3.11). Because of this, the material dredged is a cherished commodity for local authorities in the south of England. Sometimes this material is required within a port development itself, for example for reclamation purposes. Capital dredging projects occur infrequently and sporadically, can potentially have a major environmental impact, and cannot be relied upon as a reliable source of aggregates.

3.5.8   Pulverised fuel ash

There is a gradual increase in the PFA utilisation in road sub-base construction as there is in non-aggregate applications. However, difficulties in obtaining environmental approvals for use in unbound applications (eg as a fill material) has prevented more extensive application (ODPM, 2001).

The elements present in PFA are the same as those associated with the plant life that formed the coal deposits and are themselves natural materials. The properties of PFA can be summarised as follows:

- the majority of the ash is present as an alumino-silicate glass
- most elements are present in very small quantities and are largely entrained in the glassy material
- typically less than 2 per cent of the PFA is water-soluble; calcium and sulphate constitute the majority of the water-soluble fraction. There are smaller amounts of sodium, potassium and, in low pH leachate, magnesium
- the pH is mainly determined by the water-soluble calcium and sulphate
- the water-soluble fraction, though small, can be sufficient to produce a pH above 11.5, but dilution can rapidly reduce the water-soluble fraction and therefore the pH.

PFA is naturally of low permeability and in many applications is used with a binder, eg cement, lime. Its low permeability makes it difficult, if not impossible, both to saturate and to extract the water-soluble fractions. When PFA is used in road construction, or tested in a permeameter, an almost impermeable bed is created.

In an actual deposit, the very low permeability produces very small throughput of water. This, combined with the high alkalinity and very low solubility, has been found to give no problems due to leaching in practice.

Pulverised fuel ash can be used as a lightweight fill for embankment construction. It is relatively easy to place with compaction by vibrating rollers or footpath compactors, giving an optimum moisture content of approximately 19 per cent, with 10 per cent air voids. The use of grid reinforcement resistant to corrosion is recommended because of PFA's fine structure. Care must also be taken in drainage, as PFA is particularly sensitive to the effects of uncontrolled water – a particular concern for river and flood protection embankments. Some PFA materials display pozzolanic properties, but this cannot be relied upon in the design (Barber *et al*, 1972).

The construction industry does not recommend the use of unbound PFA below the water table; for such applications cement/PFA grout or concrete is required. The recommended design of PFA embankments requires a capillary break/drainage layer beneath the PFA and protection in some manner to all exposed surfaces. However, a CIRIA Report 167 on the use of industrial by-products in road construction (Baldwin *et al*, 1997) concluded that it was reasonable to consider the *in-situ* leaching behaviour of PFA as similar to materials such as recycled brick rubble and crushed concrete.

3.5.9 Glass cullet

Glass cullet may be pulverised into a sand-like product, for which there are limited applications as fill material and for drainage.

The amount of glass used as aggregate has rapidly increased over the last few years and it is estimated that potentially it could amount to several million tonnes. There are examples of its use for beach recharge in the USA (eg Fort Bragg, California). However, waste glass in the UK commands a prohibitively high market value, as demand outstrips supply, and it may not at present be available in sufficient quantities to satisfy bulk applications in coastal and river engineering projects. It also requires some treatment before use (ODPM, 2001).

3.5.10 Spent foundry sand

Foundry sand had been used as structural fill in the USA for many years. When new environmental laws were introduced this practice ceased. Nevertheless, Ham *et al* (1990) concluded that foundry wastes from ferrous foundries are generally non-hazardous. This finding has allowed several states to issue exemptions or blanket policies for the beneficial reuse of foundry wastes in structural fill applications. In Ohio, recycled foundry sand is being recommended for use as "select granular backfill" by the Department of Transportation after new State standards took effect in the mid-1990s. Over the past few years the Ohio Turnpike Commission has used more than 158 000 t in road embankment construction projects (<http://www.foundryrecycling.org>). Most states, however, place restrictions on locations of such applications and require the material to be encapsulated in some way.

3.6 SUMMARY

All of the materials considered have the potential for further application in coastal and river engineering, some quite extensively, but information regarding use in most cases is scarce and poorly documented. British Standards for the use of these materials for such structures and applications is patchy and are in any case being replaced by European Standards.

Table 3.6 gives some suggestions for the ways in which alternative materials might be used for common coastal/river engineering scheme elements. The information presented in Table 3.6 is based on a preliminary assessment and should not be seen as definitive or prescriptive. Particular applications will depend on availability and cost.

Table 3.6 *Suitability of alternative materials for common coastal/river engineering scheme elements*

Alternative materials	Concrete seawalls* — Fill	Concrete seawalls* — Prom surface	Coastal revetment/ beach control structures* — Core/under-layers	Gabions, eg walls, groynes	Geobags	Beach recharge	Cliff drains	Saltmarsh cheniers	Embankments — Fill	Embankments — Revetment	Flood walls	Riverbed protection *
Recycled aggregates												
Granular materials	✓	C, B	✓	✓	✓	✓	✓	✓	✓	C, B	C	✓
Maintenance dredgings (muddy)	✗	✗	✗	✗	✗	✓	✗	✓	✗	✗	✗	✗
Capital dredgings (sand, gravel)	✓	C, B	✓	✓	✓	✓	✓	✓	✓	C, B	C	✓
Spent railway ballast	✓	✗	✓	✓	✓	✓	✓	✓	✓	✓	✓	✓
Recycled concrete rubble	✓	B	✓	✓	✗	✓	✓	✓	✓	✓	✓	✓
Kerbstones	✗	✗	✓	✓	✗	✗	✗	✗	✗	✓	✓	✓
Railway sleepers	✗	✗	✗	✗	✗	✗	✗	✗	✗	✓	✓	✓
Secondary aggregates												
Burnt colliery spoil	U, C	C	✓	✗	?	✓	✓	✓	U, C	C, B	C	C, B
Unburnt colliery spoil	U, C	C	✓	✗	?	✗	✗	✗	U, C	C	C	C
Steel slag (EAF, BOF)	✓	C, B	✓	✗	?	✓	✓	✓	✓	C, B	C	C, B
Blast-furnace slag	✓	C, B	✓	✗	?	✓	✓	✓	✓	C, B	C	C, B
Furnace-bottom ash (FBA)	C	C	✓	✗	?	✗	✗	✗	C	C	C	C
China clay sand	✓	C, B	✓	✗	✓	✓	✗	✓	✓	C, B	C	C, B
Slate aggregate	✓	C, B	✓	✗	✓	✓	✓	✓	✓	C, B	C	C, B
Foundry sand	✓	C, B	?	✗	✓	✓	✗	✓	✓	C, B	C	C, B
Recycled glass	✓	C, B	?	✗	?	✓	✓	✓	✓	C, B	C	C, B
Incinerator bottom ash (IBA)	✓	C, B	?	✗	?	✗	✓	✗	✓	C, B	C	C, B
Recycled tyres (in bales etc)	✓	✗	✓	✓	✗	?	?	?	✓	✓	✗	?
Pulverised fuel ash (PFA)	✓	C, B	✓	✗	✗	✗	✗	✗	✓	C, B	C	C, B

✓ generally suitable; C: suitable if bound in concrete; B: suitable if bound in bitumen/asphalt; H: suitable if hydraulically bound; U: suitable if unbound; X: unsuitable.

* Large-mass units or structures are required for most exposed locations, eg concrete armour units can be made using secondary aggregates. For sheltered sites, some recovered C&D waste may be suitable, eg concrete railway sleepers or kerbstones.

Engineers or designers will need to examine actual material properties in relation to their requirements for particular scheme elements. Case studies of past uses of alternative materials can be found on the AggRegain website <http://www.aggregain.org.uk>. This tool also helps specifiers and buyers choose the right aggregate for the application and to download detailed technical notes and purchase orders.

Other useful sources of information include the CIRIA database of construction-related recycling sites in the UK that accept and/or sell materials (at <http://www.ciria.org/recycling>) and the BREMAP database, <http://www.bre.co.uk/services/BREMAP.html>, which gives the location of waste management facilities and materials/products within a region or radius of a chosen distance based upon a specified site postcode.

4   Overcoming the barriers

4.1   INTRODUCTION

The importance of using secondary and recycled aggregates in construction is reflected in targets being set by both the ODPM and the Environment Agency for their use. This desire has resulted, for example, in considerable research, guidance and use of such materials in highway engineering (in sub-bases and embankments). A recent paper by Elliott, Ghazireh and Cole (2003) reproduces guidance on the suitability of a wide range of secondary and recycled aggregates for highway construction, indicating how they may be used.

There are, however, greater challenges to using such materials in coastal and river engineering. For example, a flood embankment will normally need to be better able to prevent seepage of water than a road embankment and will require protection against erosion by currents or waves. Even with such protection, the possibility of failure of a flood embankment remains, with the consequence of materials used in its construction finding their way directly into the river or sea.

Whereas some barriers to using secondary and recycled aggregates are identical to those for other construction projects, for example availability and economics, extra concerns apply in coastal and river engineering projects. These include aesthetics and the release of small particles into the water, causing turbidity and siltation. Nevertheless, there is a growing list of successful applications, exemplified by the case histories presented in this book. With further education, dissemination and research, it should be possible to reduce the use of primary aggregates further in the future.

This chapter considers the barriers to using secondary and recycled aggregate. To this end, a workshop was held (see Appendix 5) as part of the project, involving a range of stakeholders. Solutions to the identified barriers were also discussed with stakeholders at the workshop. The workshop discussions and findings have been used extensively in writing this chapter, which considers the barriers and makes recommendations on ways of overcoming them under the following main headings:

- perceived quality and specifications of materials
- availability, transport and economics
- environmental concerns
- policies, legislation and regulations
- perceptions of risks and liability.

The final section of the chapter summarises the recommendations on ways to overcome barriers to increased use of alternatives to primary aggregates.

4.2   PERCEIVED QUALITY AND SPECIFICATIONS OF ALTERNATIVE MATERIALS

Existing guidance for the design of river and coastal engineering structures is rather patchy, but for some scheme elements it can be particularly prescriptive, for example specifying the strength and durability qualities of concrete or armour rock, or the grain size distribution of sediments to be supplied to beaches. A requirement for such high-quality materials can be justified in locations such as the Blackpool and Bournemouth seafronts, yet there are many other situations where the hydraulic environment and amenity usage are less demanding. Applying to these cases a rigorous specification for colour or shape of aggregate particles, for example, can impose unnecessary restrictions on opportunities to use secondary or recycled materials.

There are three main ways to overcome the unwarranted exclusion of secondary or recycled construction materials on the basis of their quality and consistency.

1   Improving the consistency and quality control of alternative materials.

2   Adjusting the design of engineering schemes to allow the use of alternative materials.

3   Use of performance specifications.

On the first of these, many credible processors/recyclers can guarantee the quality and consistency of their products, but designers and contractors do not always know this. There can also be a lack of consistency in quality between producers. The quality of the material will often be dependent on the quality control at source, and this is particularly the case for recycled materials. The OPDM (2001) concluded in its report that the key to increasing the proportion of hard C&D waste recycled as aggregate is to improve on-site separation of these materials from soil and other potentially deleterious materials so as to make them more acceptable to the recycling industry.

Concerns about the possible inclusion of contaminants, organic waste or plastic debris may prejudice the case for using alternative materials, particularly since the designer of a scheme will be responsible for any consequential adverse effects on the environment. The perception that the specifications do not exist for, or include, secondary or recycled materials is changing. Elliott, Ghazireh and Cole (2003) provide various information sources and specifications for such materials in the context of highway construction, and there is a need for similar information and specifications for coastal and river engineering projects.

Another way to reduce the use of primary aggregates is to design works taking into account the properties of alternative materials, once these can be specified and supplied reliably. Such design changes may include, for example, increasing the dimensions of structures or slackening slopes. Note that this may result in the need to use a somewhat larger volume of secondary or recycled materials than of primary aggregates, and/or the need to carry out maintenance more frequently.

In summary, therefore, designers could adjust designs of schemes to make use of secondary or recycled aggregates if those materials were well specified and of consistent quality. Those interested in selling such materials need to disseminate information on their products more effectively and to ensure that negative perceptions are overcome by ensuring high quality in the materials they provide.

4.3   AVAILABILITY, TRANSPORT AND ECONOMICS

Uncertainty regarding the availability of secondary and recycled aggregates in suitable quantities, and at the required time, acts as a barrier to the use of alternatives to primary aggregates in coastal and river engineering. While the overall quantities of unprocessed alternative materials in the UK are large, they may be distant from the location of a particular scheme and may need to be sorted before delivery.

For schemes such as the construction of embankments or beach recharge, which account for a large percentage of the primary aggregates used, the continuity and rate of delivery of the basic materials is an important factor in the timely and cost-efficient completion of such schemes.

One solution to this perceived difficulty would be to stockpile secondary or recycled aggregates, having carried out the requisite sorting and quality checks, so that such delays do not occur. Such an approach would be expensive, and risky if the demand for the materials did not reach expected levels. Also, there are restrictions on creating and maintaining such stockpiles (see Section 4.5). The rate of supply of alternatives to primary aggregates therefore seems likely to be limited to the rate at which the raw materials can be processed. While this processing rate may be reliably estimated for materials such as slate aggregate or china clay sand, it is less easy to predict the availability of recycled aggregates, such as those arising from construction or demolition schemes. Although demolition works and port dredging operations can generate large quantities of suitable material, these activities may not coincide with major engineering works that can use it.

A further potential difficulty is a lack of sufficiently detailed information on alternative materials – for example, the particle size grading. This is sometimes required at the design stage for a project, which could be years in advance.

This may be overcome by allowing design changes immediately prior to, or during, construction, so that the contractor can take advantage of the availability of materials at the time.

If the difficulties of ensuring supplies of alternative materials can be overcome, the next barrier to their usage is their cost. From the outset, it must be recognised that most river and coastal engineering projects are, and will continue to be, low-cost projects. Many recycled aggregates are already being used in higher-value applications, so there is little likelihood of these being used in the future. For other alternative materials, particularly recycled aggregates, there is scope for greater usage.

The costs of any construction materials for any scheme reflect those of obtaining the raw materials, of the processing required to produce the required consistency and quality, and of their transport to the construction site.

On the first of these issues, alternative materials have a potential advantage, reinforced recently by the introduction of the Aggregates Levy on the production of primary aggregates. While the costs of obtaining the "raw" secondary or recycled aggregates is low or nil, there are costs involved in managing them, for example waste management licences.

Both primary and alternative aggregates require processing to ensure their consistency and quality, by removing contaminants for example. These are likely to very similar for primary and secondary aggregates, but they may be higher for recycled materials such as demolition waste.

Finally, the cost of transporting secondary or recycled aggregates to the locations where river or coastal engineering works are being undertaken is an important factor in the total cost of those materials. For coastal projects, marine transport is the preferred method of delivery for large quantities of materials to the site, to avoid disruption to the local community, pollution and other adverse environmental effects. Many sources of primary aggregates, such as quarries and offshore dredging areas, can deliver materials by sea or rail, hence reducing costs and environmental damage by having much lower greenhouse gas emissions and embodied energy per tonne/tonne-kilometre when compared to delivery by road (Masters, 2001). In contrast, many of the major sources of alternative materials are far from the areas of demand, and only rarely are there any alternatives to using road transport.

There may be concerns about further potential costs associated with the use of alternative material, for example arising from delays in delivery, or associated with environmental assessments or licensing of temporary stockpiles.

At present, therefore, the economic case for using alternatives to primary aggregates in coastal and river engineering is not convincing, largely because of the costs of transport. While there are environmental benefits to using such materials, and targets have been set to encourage their use, their uptake will increase only if there are significant savings in the costs of a scheme compared with using primary materials.

For flood defence or coastal protection schemes, for example, the need to avoid unnecessary costs is crucial when making a case for grant aid and in setting the priority for a scheme. While some account is taken of environmental issues in this appraisal process, no tangible credit is given for using alternative materials. At present, therefore, these materials need to compete with primary aggregates on cost as well as to overcome the other barriers discussed in this chapter.

One obvious advance would be to reduce the costs associated with the processing and transport of alternative materials, and some progress has already been made in this respect. Facilities to deliver china clay sand by sea are being planned in Cornwall, and the provision of similar facilities for transport by sea or rail would be helpful in improving the uptake of other alternative materials. The National Assembly for Wales (Ove Arup, 2001) concluded that rail transport was the only sustainable option for moving secondary aggregates derived from slate quarrying. It found, however, that the existing funding framework would not be able to deliver the major investment required for rail freight improvements. Such funding would be possible only if modifications were made to the rail freight grant system.

Costs could be reduced further by, for example, reducing the amount of processing or by the development of less costly processing technology.

If a different and wider-ranging approach to assessing the economics of using alternative materials were to be adopted, then this may also reduce the usage of primary materials. This might involve accounting for the environmental consequences of producing primary aggregates and of disposing of secondary or recycled aggregates. Life-cycle analysis and cost-benefit analysis are not always applicable when assessing the use of recycled and secondary aggregates. Multi-criterion analysis may be a more appropriate assessment method, as this takes environmental issues into account. This would need to include some quantification of environmental costs and benefits.

Following this, the costs to individual schemes of using alternative aggregates could perhaps be reduced by means of subsidies to reflect the overall environmental benefits to the country from their use in preference to primary aggregates.

In summary, therefore, if costs of delivering recycled and secondary aggregates to a site can be reduced, and design engineers and contractors can be convinced that quantities and supply schedules can be met, the use of such materials will increase automatically.

4.4   ENVIRONMENTAL CONCERNS

The substitution of secondary or recycled for primary aggregates can lead to concerns about potentially greater effects on the environment. These will mainly occur at or near the construction works, but may also be associated with transport of the materials, especially if greater distances have to be covered. The sensitivity of coastal and riverine environments makes this an especially important consideration, particularly in highly sensitive environments such as lakes and slow moving water.

Potential environmental effects are considered under the following headings:

- chemical
- physical
- biological
- aesthetic and socio-economic.

4.4.1   Chemical effects

The vast majority of recycled and secondary aggregates are similar in character to primary aggregates. Even if they are subjected to erosion or abrasion and break down into smaller particles they remain inert and present no chemical threat to the environment. The paper by Elliott, Gazireh and Cole (2003) indicates that for highway construction projects few of the alternative materials considered are considered unsuitable for use even as an unbound sub-base for road construction, for example PFA, FBA and unburnt colliery spoil.

Nonetheless, concerns still arise about the potential for some alternative materials, such as tyres or IBA, to contain small quantities of contaminants such as nitrates or heavy metals. Even if the percentage of these is low, the large quantities of materials required (in embankment construction, for example) raise fears about contaminants being leached out into the water.

A previous CIRIA research project examined whether industrial by-products and reclaimed materials used in road construction and earthworks are likely to cause contamination of ground and surface water. The data obtained from this research and published in Baldwin *et al* (1997) provides a baseline appreciation of the leaching behaviour of a wide range of industrial by-products, including most of those considered within this book.

The conditions and the laboratory testing apparatus were set up to imitate leachate behaviour in road construction layers. The results from this study presently provide the best knowledge to date of comparative leachate behaviour from these materials. Further research is needed to provide information on how various alternatives to primary aggregates will perform in river or coastal engineering applications, whether loose or bound, eg in concrete or bitumen. Particular consideration will need to be given to the effects of salt water and mechanical abrasion that would be encountered in coastal applications.

In summary, while the potential for chemical effects on the environment of all construction materials always needs to be considered, the great majority of the recycled and secondary aggregates considered here are inert. With appropriate use, they will pose no greater threat to the environment than primary aggregates.

4.4.2 Physical effects

The use of secondary or recycled aggregates in engineering schemes may have a variety of effects on the physical environment. For example, alterations to the dimensions of a structure to allow the use of such materials may enlarge its footprint, hence increasing the area affected by their construction. If these materials contain significant amounts of fine-grained particles, or could degrade to produce such amounts, then the dispersion and subsequent settlement of these particles can affect water clarity and cause changes in the sedimentary or morphological characteristics of the river or seabed.

One example of this is the possibility of a change to the permeability, and hence the gradient, of a beach. This in turn could cause an increase in the rate of sediment transport along a coastline, leading to changes in beach levels in front of seawalls or could affect existing patterns of shoreline change.

Such concerns, however, affect all schemes that involve the addition of non-native materials, whether primary aggregates or not, and each case needs to be assessed in the light of the individual circumstances. Various methods of mitigating physical effects can be applied, such as by reducing the rate of introduction of fresh materials. The use of recycled or secondary aggregates should not be restricted by such concerns.

4.4.3 Biological effects

Rivers, coastal waters and the adjacent land margins are often of great importance as habitats, supporting a wide variety of plants and animals. Consequently, it is normal to undertake a careful appraisal of the effects on biology of any engineering schemes in such areas. The chemical and physical effects discussed above may also create extra biological impacts. The potential biological impact of using alternatives to primary aggregates will need to be assessed, therefore.

It is important to note that the environmental characteristics of a location for proposed river and coastal engineering work could be unique. Although some general guidance is possible on the acceptability of such schemes, whether using primary or alternative materials, each project proposal has to be judged on its particular circumstances. For example, concerns about the effect on water quality will be greater in areas where the water is particularly clear and for rivers used by salmon and trout for breeding.

Another example is the potential effect on commercial shellfish beds. If an engineering scheme introduces into an estuary large amounts of fine-grained material, the shellfish may be smothered. Fill operations using materials, whether primary aggregates or not, containing a large percentage of small particles would be a cause for concern, there-fore. This may mean that the construction activities have to be regulated, perhaps to limit operations to particular states of the tide, so ensuring dispersal away from the shellfish beds. Regulators (see Section 4.5.3) may further require the implementation of measures to minimise the effect on water quality, such as the use of silt curtains, geotextile membranes or even the creation of an impervious bund to retain the fine-grained materials.

Where a scheme is proposed within or close to an area with conservation designations, eg SSSI, SPA, SAC, then greater caution will be needed. In general, however, the approach to assessing the effects on biology of coastal and river engineering schemes will be broadly the same whichever construction materials are used. Extra weight will need to be given in such assessments to the consideration of possible changes in chemical and physical conditions to allay concerns about the use of alternatives to primary aggregates.

To assist in this type of assessment, greater knowledge is needed from pilot studies and research on the effects of recycled and secondary materials on the natural habitat. Some guidance on long-term effects can be obtained from areas where such materials have been introduced into coastal or riverine environments in the past, either as part of an engineering scheme or simply as waste disposal. Many beaches around the UK have been augmented by the disposal of mining or quarrying waste, and the effects on the local natural environment could be assessed. Comparative biological assessments could also be made in cases where alternative materials have been used in engineering projects. While such studies are being carried out, it may be preferable to avoid the use of some alternative aggregates in or near SSSIs, SPAs and other designated areas.

The natural environment may also gain from increased use of secondary and recycled materials. For example, several experimental schemes to restore inter-tidal habitats in Essex have been carried out using dredged material from navigation channels or recycled materials (redundant lighters). These schemes would not have been affordable if primary aggregates or materials had been specified.

4.4.4 Aesthetic and socio-economic issues

The aesthetics of coastal and river environments can be very important to the area, particularly if tourism, recreation and related activities depend upon it. There is already some resistance to the use of novel types of structures in river and coastal engineering schemes, for example rock groynes, because of the perceived aesthetic effects, and in areas designated because of their scenic qualities there are often guide-lines on appropriate forms of construction. These sensitivities may limit the use of secondary and recycled construction materials in coastal and river engineering, particularly the use of waste tyres or C&D waste if these materials become visible. For most applications and alternative materials, however, any aesthetic impact may be more perceived than actual. In Cornwall, many beaches formed largely of mine waste are regarded not only as acceptable but also as an asset to the landscape and tourism.

In general, the public supports recycling initiatives and schemes that protect, or reduce impacts on, the environment. However, the language often used to label alternative construction materials – eg "waste", "slag", "scrap" – may give the impression that there will be dangers if these were placed in areas of importance to wildlife, recreation and tourism. These terms can lead to the view that the engineering schemes themselves are inferior to those where primary aggregates are used, and they certainly do not help to convince the public that there are potential advantages to using them.

Improving the terminology, and disseminating information about successful schemes using alternative materials elsewhere in the UK, should address such concerns. The positive aspects of overall benefits to the environment in adopting more sustainable design and construction methods also need to be emphasised. If these two measures are adopted, together with appropriate design techniques, public anxiety about use of secondary or recycled aggregates could be allayed or even reversed, producing a positive view of the reduction of use of scarce primary aggregates.

4.5 REGULATION, LEGISLATION AND POLICIES

Certain regulations, policies and legislative measures relating to waste materials, environmental protection and the funding and approval of river and coastal engineering schemes act as barriers to the usage of alternatives to primary aggregates. The most influential of these are discussed below.

4.5.1   Waste classification

Classification of waste materials (Waste Management Licensing Regulations 1994) acts as a barrier to the use of alternatives to primary aggregates. For example, most mine and quarry wastes are exempt from Landfill Tax, so there is no incentive to utilise them more profitably. The labelling of these materials as "waste" further detracts from their perceived attractiveness as a potential replacement for primary aggregates. Most of the materials under consideration here are referred to, and are legally defined as, "waste". The regulations and controls governing storage, movement, use, disposal and taxation of different types of waste material feature strongly in any consideration for potential utilisation.

Article 1 of the EC Framework Directive on Waste (75/442/EEC as amended by 91/156/EEC) defines "waste" as any substance or object which the holder discards or intends to or is required to discard. This definition has been transposed into UK law by Regulation 1 of the Waste Management Licensing Regulations 1994. Schedule 4, Part II of the Regulations lists those material types that are classed as waste. Category 16 of Pt II, Schedule 4 of the Regulation, however, states that "any materials, substances or products which are not mentioned in the [other] categories can be waste when discarded" – therefore anything can be "waste".

So the question as to whether a material is a waste or not hinges on the term "intention to discard". Deciding if something has been discarded is based on whether it is no longer part of the normal commercial cycle or chain of utility.

Guidance was given on this in Annex 2 to Circular 11/94 (DoE, 1994), which states:

> ...substances or objects which may not be waste include:
>
> - those which are sold or given away and can be used in their present form or in the same way as any other raw material without being subjected to a "specialised recovery operation"
> - those which are put to beneficial use by their producer.

The European Court has determined that material should be regarded as "discarded" even if it is subjected to a disposal or recovery operation. Under European case law it is clear that even if a contract is in place for the sale and subsequent reuse of arisings, it may still be classified as directive waste if the holder's intention is to discard (Stubbs, 1998).

The Environmental Protection Act 1990 Part II and the Waste Management Regulations 1994 stipulate that a waste management licence is required by anyone who deposits, recovers or disposes of household, commercial or industrial waste (called "controlled wastes"). The Environment Agency in England and Wales and the Scottish Environmental Protection Agency (SEPA) in Scotland are the relevant authorities with duties under the Regulations to administer the waste management licensing system.

The Environment Agency's classification of secondary/recycled materials as waste often results in the need for a waste management licence, which can incur further costs on the scheme. In relation to this are the stockpiling and storage restrictions that apply to classified wastes (paragraphs 13,15, 17 and 24, Schedule 3 of Waste Management Licensing Regulations 1994) (see Appendix 4). These can inhibit the amounts of materials that can be processed and stored for future use, effectively preventing the accumulation over appropriate timescales of sufficient amounts of material for use in coastal or river engineering schemes.

If these licensing issues are addressed, however, this present barrier to the use of secondary and recycled aggregates can largely be overcome.

4.5.2   Water Framework Directive

The Water Framework Directive (2000/60/EC) came into force in December 2000. It is aimed at raising coastal and inland water quality and aquatic ecosystem health across Europe to a "high" status by 2015. It requires member states to take action to protect the morphology and the ecological status of watercourses. It is not yet clear what impact the directive will have on the construction of river and coastal defences, or on such activities as maintenance dredging.

New obligations under this directive, particularly the stringent specifications for pollutants from diffuse sources, may act against the use of alternatives to primary aggregates in water environments. This may be more relevant to certain alternative aggregates than others, but until the leachate levels for specific materials in water environments are known these obligations might act as barriers for all materials.

Since the overall purpose of this directive is to improve the "water environment", however, it may be possible to argue for the wider environmental benefits of reducing primary aggregates in coastal and river engineering schemes.

4.5.3   Conditions on licences and consents

Conditions placed on planning permissions for works, such as specified finishes for schemes, can inhibit the use of secondary and recycled aggregates in coastal and river schemes. Often the people responsible for writing conditions on licences do not have the technical expertise to take into account the implications of the restrictions they place on materials, which can adversely affect the use of alternative materials. Stringent specifications under Food and Environmental Protection Act 1985 (FEPA) licences relating to the use of materials, particularly in environmentally sensitive areas (eg those designated as SPA or SAC) limit the range of materials that can be used in engineering schemes, for example to ensure that the character of the waters or seabed are not altered. FEPA licence applications are considered on a case-by-case basis, reflecting the possible differences in each location. This can result in uncertainties at the design stage regarding which materials may be acceptable. To avoid delays later on, therefore, there is a tendency for engineers to opt for more "natural" materials, ie primary aggregates.

As experience of using secondary and recycled aggregates increases, and dissemination of the successes and benefits of such schemes takes place, regulatory authorities will become less cautious about further schemes using these materials.

4.5.4   Government policy

There is currently no positive incentive to use alternative aggregates in coastal and river engineering schemes. If there were economic gains through lower costs of such materials or better recognition for such schemes and those engineering them, the uptake of secondary and recycled aggregates might improve.

It is considered that central, regional and local government offer too little support for the use of these materials in such schemes. For example, the Defra appraisal process for coastal and river schemes usually dismisses any uncertain solutions at an early stage. Schemes involving alternatives to primary aggregates about which doubts exist over, say, environmental impacts or the supply of construction materials are likely to be discounted in preference to "safer" designs using tried and tested methods and materials. Defra's requirement for publicly funded schemes to be assessed on a cost-benefit basis, as noted in Section 4.3, also acts as a barrier, because these assessments rarely prove favourable for recycled or secondary materials.

Setting targets to encourage the use of secondary or recycled aggregates is regarded as a better way of encouraging their use than providing subsidies.

4.6   PERCEPTION OF RISKS AND LIABILITY

To date the use of alternatives to primary aggregates in coastal and river engineering has, in general, been insufficiently researched, and existing applications are poorly publicised and documented. This has led to a perception of extra risk in the design and construction of such schemes if alternative materials are used, although it is difficult to quantify any increase in the risk. This perception is based on the supposition that extra risks could arise from the quality, availability or environmental effects of such materials. Should problems arise, during or after construction of a scheme, as a consequence of choosing to use alternative materials, then the liability of the engineer may be increased. This perception acts as a barrier to use at the design stage, encouraging the engineer to opt for tried and tested materials as a safer approach, bearing in mind the burden of professional indemnity insurance.

Uncertainty about the materials' performance – ie their consistency and durability – is the main factor in the perceived increase in risk associated with their use. Until more is known about these materials and how they perform in comparison with primary materials, this uncertainty will remain, whether the risk is actual or perceived. The risk associated with each material is based on perception rather than fact; this situation will not change until the real risk relative to each material has been determined.

Unfortunately, the perceived risk associated with the use of alternative materials outweighs any expected benefits to the environment of a successful scheme. In addition, any cost advantages are rarely sufficient to affect a decision to use these materials instead of primary aggregates. Engineers, clients and funding bodies are thus less likely to choose to use alternatives to primary aggregates in coastal and river engineering.

This barrier can be reduced by addressing two key issues. First, specific, detailed facts about the quality, availability and environmental acceptability of alternative materials can alter the perception of risks. These issues have been discussed individually above, but it is probably fair to say that the most convincing demonstration of the suitability of alternative materials will come from the proven success of previous schemes similar to the one under consideration.

Second, most river and coastal engineering projects are built with public funding, and reducing the use of primary aggregates is in the best interests of the country. It seems logical, therefore, that the clients and funders of such schemes should show tangible support for the use of alternative materials. Consideration needs to be given to these public bodies accepting any extra risks arising from the use of such materials, thus demonstrating their support for the sustainable use of resources and reducing the concerns of, and risks to, designers and engineers in using them.

In summary, it is likely that it is the perception rather than the reality of extra risks in using secondary or recycled aggregates that is a barrier to their use. If funding agencies accept part of the risk in schemes using alternative materials, and set targets for their usage, progress will be made towards their ever greater uptake.

RECOMMENDATIONS ON OVERCOMING BARRIERS

There are some genuine concerns about the use of alternatives to primary aggregates in coastal and river engineering projects. These concerns can prevent increased usage of those alternatives. However, some, apparently false, perceptions about these materials remain, despite the considerable advances made in recent years in their quality, consistency and availability.

To bridge this gap between perception and reality, a better interchange of information is required, involving all parties. In many cases, there is no clear specification of the type of materials required for even relatively common forms of coastal and river engineering – eg the construction of flood embankments. If such specifications were available, then those supplying secondary or recycled aggregates would be in a far better position to satisfy market demands and provide assurance on the quality of their products. There would also be merit in active marketing of such materials by suppliers, with assistance from public-sector bodies (Defra, the Environment Agency and local authorities, for example), particularly by promoting successful applications.

At the same time, much coastal and river engineering has been carried out on the basis of making best use of cheap and readily available materials, modifying designs to match local conditions and available funds. If those planning and designing schemes had better information on the types, availability and location of recycled or secondary aggregates, they may well be able to change the type of a structure, or adjust its dimensions, to make best use of these materials. This approach to overcoming reluctance to use alternatives to primary aggregates will probably best be achieved by a mixture of independent research and dissemination, particularly related to successful uses of such materials in coastal and river engineering schemes. The difficulty experienced in finding information on such case histories during this project indicates the difficulties a design engineer would face in obtaining relevant information for a specific scheme. It is therefore likely that some new pilot projects will be needed as well as better dissemination of past schemes built using recycled or secondary aggregates.

When a single engineering scheme is considered in isolation, the costs of construction materials and the potential risks of innovative construction methods will always be important to a design engineer. The wider benefits to the environment, and to the sustainable use of resources, are not presently reflected in any tangible way in the evaluation or funding of such a scheme. Some progress on reducing the costs of alternative materials may be possible, particularly by investment in cheaper and more environment-friendly methods of delivery, such as by sea or rail. In addition, there is a case for "positive discrimination" in favour of alternative materials, by setting targets for their use or by sharing any extra risks in their usage.

It is likely that the public will have to be persuaded that the use of some alternatives to primary aggregates is safe, has advantages, is compatible with local needs for amenity, recreation and aesthetics, and provides a net benefit to the environment. Care should be taken to use non-technical language and to avoid prejudicial terms such as "slag", "spoil", "waste" etc when describing secondary or recycled aggregates. A general raising of awareness of the benefits to the country of using such materials is worthwhile, for example through the print and broadcast media. However the greatest need for education is at a local level, for each specific engineering scheme. Effective consultation and engagement with a local community at an early stage, and throughout the construction process, will assist in gaining public support for a scheme. This could be achieved via a forum for discussion with all stakeholders, public exhibitions and education through schools. Local authority awards or a mark of recognition from other interested organisations (conservation bodies, for example) would also help establish public confidence in the use of recycled and secondary aggregates.

5 Conclusions and guidance

5.1 INTRODUCTION

The conclusions and initial guidelines set out in this chapter are based on the premises that:

- reducing the quantity of primary aggregates used in coastal and river engineering will help to conserve natural resources
- replacing such primary materials with secondary aggregates (ie by-products of other industries) or with construction and demolition waste would help reduce the effects of those activities on the environment, for example by reducing the demands for landfill sites
- there may be a further benefit in reducing the costs of construction of flood defence schemes, or other engineering works along coastlines or rivers, without adverse effects on the environment.

The use of such alternative construction materials in this particular field of civil engineering is particularly challenging because the aquatic and marine environments are both difficult to operate in and environmentally sensitive. In addition, there are often concerns about the aesthetic appearance of any works and constraints on available funding, for example for flood prevention schemes.

If alternatives to the use of primary construction materials can be found for these applications, then it is likely that other civil engineering projects will benefit from the same techniques.

5.2 CONCLUSIONS

Past uses of such alternative materials in river and coastal engineering projects in the UK, and indeed around the world, have apparently been sparse and largely restricted to small-scale projects, making use of materials available locally. The driving forces for such past schemes have been the need to keep costs as low as possible and, in some cases, to dispose of unwanted waste.

This study has considered a wide range of alternative materials and examined their potential and suitability for use in river and coastal engineering projects. The decision to use any alternative materials will depend upon a wide range of issues including:

- delivery costs
- availability
- durability
- chemical composition/contaminants
- aesthetic qualities
- percentage of fines
- the amount of processing required before use.

This book shows that many secondary and recycled aggregates are technically suitable. Three main types of alternative materials are likely to be available in sufficient quantities and at reasonable cost, namely:

- by-products of quarrying and mining (mostly china clay waste and slate waste)
- C&D waste
- scrap tyres.

The main opportunities for the potential use of these materials in the context of river and coastal engineering are outlined below.

5.2.1 By-products from mineral extraction operations

- Large stockpiles of available material; better bulk transportation/shipping links will drive down transport costs and improve their economic viability
- existing examples of use for coastal and river engineering indicate their suitability for schemes (eg beach recharge)
- quality issues need to be addressed, extraction of fine constituents, potential contaminants, durability/hardness etc
- should only be considered for applications such as beach-recharge of non-aesthetically sensitive beaches.

5.2.2 Construction and demolition waste including clean excavation spoil

- Potential for localised stockpiling of material for specific small-scale remedial projects and subsequent use in "little and often" maintenance programmes
- a lesson learnt from past experience is that suitable crushing, grading and screening treatment is required to ensure the elimination of unsuitable materials and the consistency of quality needed for the desired application
- the opportunity to stockpile and reuse certain "complete concrete" forms (or large stone blocks); for instance, railway sleepers and other prefabricated concrete blocks should be considered for future applications in consultation with coastal engineers.

5.2.3 Scrap tyres

- Increasing availability of tyres and pilot project findings are likely to result in increased utilisation
- unique material qualities may allow for innovative design specifications (such as use in low-density bulk fill applications on soft ground)
- may develop as an important sustainable use of end-of-life tyres, especially in the many UK flood embankments that will require reconstruction or upgrading in the near future.

5.2.4 Summary

Some barriers exist to the incorporation any type of secondary or recycled aggregates in coastal and river engineering schemes. At present, it is probably fair to conclude that any potential cost-savings made in using these materials for such schemes are out-weighed by concerns about their availability, perceptions of and uncertainty over the potential for environmental damage, and the long-term durability of the resulting works. The lack of examples of previous successful uses of such materials adds to these concerns. Consequently, designers are likely to adopt a "better safe than sorry" attitude.

To make a significant reduction in the use of primary aggregates, the possible uses of secondary aggregates or C&D waste outlined in the following sections should be considered.

5.3 GUIDANCE FOR COASTAL MANAGEMENT PLANNING AND FUNDING BODIES

Coastal management planning and funding bodies are recommended to:

- provide incentives to increase attractiveness of alternative materials, eg in a "priority scoring" system

- arrange facilities for sorting, quality-checking and stockpiling of suitable materials

- move towards smaller, more frequent management schemes requiring shorter service lives

- clarify sustainability definitions when applied to coastal or river engineering schemes.

5.4 GUIDANCE FOR COASTAL ENGINEERS/CONSULTANTS

To make a significant reduction in the use of primary aggregates, then the following possible uses of secondary aggregates or C&D waste should be considered. These major applications have been selected on the basis of the amounts of money being spent annually in the UK on the replacement of existing coastal and flood defence works. Note that the advice given here is necessarily general in nature. As with all schemes, the location and the particular characteristics of the site will have a considerable influence on the types of work considered, and hence on the feasibility of using secondary aggregates or construction/demolition waste.

As discussed in Section 3.4, under the new EU standards many recycled and secondary materials meet the requirements for aggregates and so should be considered along with primary materials. If these materials have particular characteristics such as unusual sizing or higher percentage of fines than required, the possibility of adjustment to the standard design of a structure should be considered to accommodate their use.

The planning for maintenance and replacement of structures should include whole-life costing as part of the project process. Integrating whole-life costing will enable sustainability, environmental impact and the material sourcing hierarchy (see Box 5.1) to be considered early in the process. By following the hierarchy shown in Box 5.1, the environmental impacts of project options can be reduced along with the costs.

Box 5.1 *Hierarchy of material sourcing options (after Masters, 2001)*

1	Suitable materials available on site from a previous scheme or structure.
2	Locally sourced alternative materials appropriate to fulfil the functions identified in the functional analysis of the project.
3	Alternative materials from farther afield that can be delivered to site predominantly by sea or rail or locally sourced primary materials.
4	Alternative materials transported from farther afield by road or primary materials transported from farther afield predominantly by sea or rail.
5	Primary materials transported from farther afield by road.

A useful detailed explanation of whole-life costing is given in *Whole life costs and project procurement in port, coastal and fluvial engineering* (Simm and Masters, 2003).

5.4.1   Beach recharge

These schemes are major users of primary aggregates, although providing sand or gravel for these schemes does not presently require any payments under the Aggregates Levy. For large schemes close to the sources of secondary aggregates, then consideration should be given to the practicality of using china clay sand or slate aggregate, from Cornwall and North Wales respectively.

For smaller schemes, such as those involving frequent top-up operations, trickle-charging or slow-release methods should be developed, using material as it becomes available. In these cases, it should be possible to use excavation spoil, and perhaps demolition waste, from local construction projects. Precedents for such operations have been set, for example using dredged material from navigation channels or rivers, or using hoggin from the excavation of foundations. These operations are more likely to be acceptable on shingle beaches away from the main coastal resorts and in areas where a localised, short-term increase in water turbidity will not have a significant effect on marine life. In Queensland, sand excavated from back-shore areas as part of the construction of seafront properties was placed on to the beach.

The use of alternative aggregate may be constrained where beaches lie within sites designated nationally or internationally for their nature conservation importance or landscape quality. Certain birds such as the little tern require a specific pebble size range for their nesting habitat. The vegetation found on some shingle beaches is of great importance from a conservation viewpoint, and recharge with stones of a different size or lithology to the indigenous beach material, with primary aggregates or alternative materials, might therefore be unwelcome.

5.4.2   Armour stone

Consideration should be given to the use of recycled concrete rubble, masonry, kerbstones and similar material as an under-layer for larger armour stone (rock or concrete armour units) in breakwaters, rock groynes and coastal/riverbank revetments. These materials have been used in the past, utilising concrete from defences that have been demolished as part of the same scheme, or as temporary protection after damage to beaches or defences. However, it should also be possible to bring in recycled concrete rubble or similar material from elsewhere as part of capital defence scheme, thereby reducing the need for quarried rock.

Options for reducing the volume of large rocks (or concrete armour units) used as the surface layer of structures built along exposed coasts are few. It is worth considering the use of concrete armour units made from recycled materials for this purpose, however. One novel idea is to use old concrete railway sleepers, perhaps joined together, to form interlocking units or large concrete blocks.

5.4.3   Seawalls and revetments

In several places around the UK coast secondary materials have been used to form seawalls or revetments, even in aggressive environments. Typically, such materials are placed into gabion baskets in suitably low-energy environments, or into more substantial crib-work structures (see Case Study A3.5) on more exposed coasts. The same materials described under the "Armour stone" heading above have been used in this manner.

5.4.4   Earth flood embankments (rivers and estuaries)

The building, upgrading or replacement of relatively simple earth flood embankments is the most expensive item in coastal and river defence in the UK, based on the estimated annual replacement costs. Counting together the embankments used in river, estuarine and coastal flood prevention, this average cost is about £63 million a year (see Simm and Masters, 2003).

Typically these are built using materials extracted from nearby, often resulting in "borrow ditches". In Norfolk, and perhaps elsewhere, such excavations are subject to minerals planning guidance and licensing, just as a sandpit or quarry would be. Reducing the amount of excavation to build flood embankments can be regarded as reducing the demand for primary construction materials, which can be achieved by, for example, using secondary aggregates or construction/demolition waste for the embankment core.

The use of C&D waste, in particular, would also cut down demand on landfill sites. In the past, embankments have been built at very low cost by allowing contractors to deliver suitable materials to the site, rather than having to pay for their disposal.

5.4.5 Revetments on riverbanks and flood embankments

As well as the expenditure on earth embankments, noted above, there are also substantial costs incurred annually in replacing or upgrading armoured or revetted embankments or channels, whether along coastlines, estuaries or rivers.

Secondary materials can be used as a substitute for primary aggregates in concrete block revetment systems, or as fill in gabion mattresses. There is also potential for the use of secondary materials in bituminous mixes as described in Section 3.4.

5.5 GUIDANCE FOR REGULATORY AND PLANNING AUTHORITIES

This guidance relates to:

- environmental protection agencies (EA, SEPA, CEFAS, EN, CCW, SNH)
- waste management authorities (EA, local/unitary authorities)
- minerals planning bodies
- local planning bodies (county/local/unitary authorities).

Any potential for reducing the costs of coastal and river engineering schemes should be a high priority for those responsible for their construction, maintenance and replacement. This includes the selection of construction materials. Choosing cheaper but acceptable resources in the form of alternative materials could, especially in large-scale projects, result in a significant reduction in project costs and usage of primary aggregates.

The best approach for increasing the uses of recycled and secondary materials under existing legal and regulatory conditions would probably be by strengthening the integration of regional minerals and waste management planning with Environment Agency (and CEFAS) guidance.

Shoreline management plans (SMPs), estuary management plans (EMPs) and catchment area management plans (CFMPs) should state a requirement to consider alternative aggregates as one of their sustainability objectives for development.

Constructive assistance should be given to test materials in pilot study applications where possible. Cost at this stage should not be a determining factor in the decision-making.

A full reappraisal of alternative materials should be made in light of the new European Standard specifications that came into effect on 1 January 2004.

5.6   GUIDANCE FOR CONTRACTORS

Contractors should be aware of the commercial advantage, both to themselves and to coastal and river engineers, of using treated C&D waste and secondary aggregate. Diverting waste to reuse reduces the cost in landfill levy to the contractor. By passing on these savings and selling the material to the coastal or river project for less than the equivalent cost of primary aggregate, costs are also reduced for the coastal engineer.

5.7   CONSTRUCTION/DEMOLITION FIRMS

Construction and demolition firms can play an important part by identifying those construction or demolition projects that may give rise to suitable material for coastal and river engineering schemes. Appropriate working practices should then be employed to ensure waste materials are adequately segregated during the demolition/construction project. Where necessary this should be agreed in consultation with recycling and reclamation firms. Implementation and proof of adherence to an industry-recognised material segregation protocol would help to persuade contractors and recycling firms that arisings are suitable.

5.8   RECYCLING/RECLAMATION FIRMS

Modern methods of crushing, screening and grading have greatly improved the quantity and quality of recycled and reclaimed C&D waste and aggregate. Recycling and reclamation firms are in a prime position to promote new uses for their products.

Contractors are in a pivotal position from which they can identify the opportunities for the use of their product, in parallel with the future supply of waste to them. If a contractor is in regular dialogue with coastal planning managers, plans can be made for suitable material provision, whether those measures include stockpiling (if provisions allow) for large-scale projects, regular supply for trickle-charge schemes, or a combination of both.

Contractors should ensure that coastal planners are aware of the potential of these resources and be able to assure continuity of supply and specified quality. In consultation with planners and engineers it may be possible in non-time-critical projects to match a schedule of work to the rate of supply rather than discount secondary aggregate or C&D waste simply because primary aggregate is more readily available. For beach trickle recharge schemes this could prove to be a dependable regular source of work for a contractor.

Recycling and reclamation firms can ensure that their products are more suitable for use by adopting an industry-accepted quality protocol.

5.9   MARINE/RIVER SPECIALISTS

Marine and river specialists could contribute in a unique way by identifying waste forms suitable for reuse without further treatment, such as kerbstones or large concrete pieces, which are particularly useful for certain applications. Notification of request for these materials given to construction and demolition firms, and to recycling and reclamation contractors, would enable them to make provision for segregation and stockpiling.

6 Future development and research needs

6.1 TARGETS AND INCENTIVES

There is a case for encouraging the use of alternatives to primary aggregates in coastal and river engineering schemes by establishing targets for their usage in publicly funded projects, and giving greater priority to schemes that meet or exceed these targets.

The continuing application of levies on primary aggregate production and on disposal to landfill should lead to alternative materials becoming more competitively priced. In this situation, market forces will naturally lead to greater production and uptake of these materials in such schemes.

With co-operation, economies of scale could result in the bulk storage and transportation of large quantities of waste materials, thereby improving the viability of segregation, separation and further processing of alternative aggregates.

A review and amendment of waste legislation and regulation is needed to engender support for more efficient resource use. A new system of regulatory definitions is needed to represent the true status of materials based upon the life-cycle of products that recognises the levels of environmental risk at each stage in their cycle.

It is acknowledged however, that a change in definitions would require a lengthy and significant revision of EU legislation and any amendment would take time to achieve.

Within existing regulation there is scope for a remodelling of guidance and attitudes to recognise the shortcomings of definitions that do not represent the true potential of recycled and secondary aggregates in coastal and river engineering schemes and applications.

Where appropriate, regulations should be applied in a way that is consistent and acts as an incentive rather than a barrier to consideration of their use.

Clearly a strategy is required if the use of alternative materials in coastal and river engineering is to be further developed. The existing sustainable waste management programme managed by Defra could be enhanced to develop further research into leachate and physical testing of alternative materials.

A technical survey of stockpiles for some materials should be considered to identify more accurately the variability, quality and status of those potential resources. This will identify those stocks that require further treatment and processing (depending on envisaged application). A database of this information along with advice made available to engineers would encourage further consideration and uptake.

6.2 TRIALS AND MONITORING OF SELECTED APPLICATIONS

An adequately funded pilot programme (possibly partly funded from the Landfill Tax Credit Scheme) using alternative materials in various coastal and river engineering structures and schemes is strongly recommended. These pilot projects should monitor and record the physical and chemical performance of the materials used. It may be appropriate for the pilot projects to be co-ordinated by a research team. In any event, the results should subsequently be assessed, reported and disseminated, demonstrating to coastal and river engineers that alternative materials can be used.

Various scheme types should be covered, including:

- beach recharge trials using china clay sand, slate aggregate and smaller quantities of recycled aggregate

- flood embankment construction using various secondary or recycled aggregates

- incorporation of recycled or recovered C&D waste, and/or secondary aggregates within revetments, breakwaters or rock groynes/sills

- employment of used tyres as fill, earth reinforcement or as part of a revetment.

6.3 EDUCATION AND DISSEMINATION

Future developments utilising alternative materials in coastal and river engineering schemes should be disseminated throughout the construction industry as widely as possible.

The results of any key modern pilots and historic case studies along with lessons learnt could be collated and presented in a published symposium as a guide to best practice in the use of alternative materials in this field. In addition, descriptions of case histories should be made available on the Internet, for example on the AggRegain website, as a source of guidance and stimulation for engineers designing coastal and river engineering schemes.

A change in engineers' and designers' perception of alternative materials from that of a waste product to that of potential resource should be engendered. There needs to be a shift in focus away from waste management to a resource management concept. Recycling or recovery of waste, production of recycled raw materials and demand for recycled products should be seen as part of a holistic system that seeks to maximise and balance the supply and demand of materials.

Designers and engineers should be encouraged to be innovative in their choice of materials. Probably many would like to be more so, but do not have the necessary materials specification information, engineering and structural guidance or track record evidence of use on which to base an informed and professional decision. Dissemination of this information should occur whenever it arises and at the earliest possible opportunity. This information should then be integrated into routine guidance on materials and design.

The training of graduate engineers should contain a basic element with regard to resource use and sustainable development. Later training should engender an attitude that looks on waste materials and secondary aggregates as a resource to be utilised whenever possible and prior to considering primary materials.

It is particularly important to publicise case studies where alternatives to primary aggregates have been used successfully. This is the best way to convince engineers, designers and other practitioners in the industry of the suitability and potential use of alternative materials in this field. The AggRegain database of alternative materials and the WRAP website should be further developed, funded and promoted to strengthen the link between engineers/designers/contractors and suppliers.

Appendices

ENVIRONMENT AGENCY COSTS FOR COASTAL AND RIVER DEFENCES

Table A1.1 *Environment Agency annual maintenance costs for river and sea defences in 1999/2000 (excluding Wales)*

Type	Defence type	Anglian Region	Midland Region	North East Region	North West Region	South West Region	Southern Region	Thames Region	TOTAL
Fluvial	Bridges			186 084	0				186 084
	Channel revetments	1 257 576	221 592	483 537	89 451	105 000	130 000	65 400	2 352 556
	Concrete/sheet piling	503 030	87 674		0	1 195 000	225 000	7625	2 018 329
	Culverts		24 859	121 336	0		20 000	0	166 245
	Ditch			8218					8218
	Flood alleviation scheme							260 128	260 128
	Flood storage & alleviation scheme (no)							80 000	80 000
	Flood wall	251 515	207 158		313 000	2 047 500			2 819 173
	Raised earth embankment	6 287 879	3 629 934	1 518 660	724 000	1 735 000	400 000	28 800	14 324 273
	Timber piling				1 288 000				1 288 000
Fluvial total		8 300 000	4 171 217	2 317 885	2 414 451	5 082 500	775 000	441 953	23 503 006
Other	Control structures: barriers/sluice/control					405 000			405 000
	Control structures: sluice/control/dropboards				33 000				33 000
	Control structures: weirs/sluice/control							0	0
	Debris, blockage clearance etc				5 000 000				5 000 000
	Flood defence gates		0	12 363			50 000	152 500	214 863
	Flood storage & alleviation scheme (no)				150 000				150 000
	Outfalls	249 000	2 346 894	1 112 000	0		578 000	45 000	4 330 894
	Pumping stations	120 000	0	480	610 000	360 000	41 000	45 000	1 176 480
	Reservoirs						90 000		90 000
	Sluices	19 500	0				300 000		319 500
	Trash screens		0					0	0
	Weirs		0	43 200				135 000	178 200
Other total		388 500	2 346 894	1 168 043	5 793 000	765 000	1 059 000	377 500	11 897 937
Sea	Beach	17 472			0	70 000			87 472
	Beach (shingle)						1 000 000		1 000 000
	Beach (shingle) and control structure						900 000		900 000
	Beach + control structure	526 750				49 000			575 750
	Counterwalls						150 000		150 000
	Dunes	0		0	0				0
	Grassed/clay embankment	0	78 854	0	92 000				170 854
	High ground		62 548						62 548
	Offshore breakwaters						20 000		20 000
	Protected embankment	1 660 000	7752	12 000	0	72 000	300 000		2 051 752
	Saltmarsh						10 000		10 000
	Seawall	1 400 000	6683	0	12 000	402 500	300 000		2 121 183
Sea total		3 604 222	155 837	12 000	104 000	593 500	2 680 000	0	7 149 559
Tidal	Composite structures							156 000	156 000
	Ditch			39 780					39 780
	Raised earth embankment	1 642 000	932 877	123 880	408 000	1 314 000	2 500 000	88 800	7 009 557
	Revetted embankment	668 000	420 730	10 230	37 000	0	366 000	99 600	1 601 560
	Seawall	248 000	47 579	8210	38 800	216 000		248 000	806 589
	Steel sheet piling	172 000	13 098	8256	0	0	165 000	187 600	545 954
	Timber piling	6000		17 850	11 200			4000	39 050
Tide total		2 736 000	1 414 284	208 206	495 000	1 530 000	3 031 000	784 000	10 198 490
Grand total		15 028 722	8 088 232	3 706 134	8 806 451	7 971 000	7 545 000	1 603 453	52 748 992

Table A1.2   *Environment Agency annual replacement costs for river and sea defences in 1999/2000 (England only)*

Type	Defence type	Anglian Region	Midland Region	North East Region	North West Region	South West Region	Southern Region	Thames Region	TOTAL
Fluvial	Bridges			104 400	122 449				226 849
	Channel revetments	1 458 333	746 100	2 403 000	10 948 925	64 913	1 083 333	436 000	17 140 604
	Concrete/sheet piling	1 200 000	966 288		1 181 250	8 767 309	1 950 000	2 541 667	16 606 514
	Culverts		146 940	366 667	77 613		333 333	990 000	1 914 553
	Ditch			3333					3333
	Flood alleviation scheme							2 153 667	2 153 667
	Flood storage & alleviation scheme (no)							66 667	66 667
	Flood wall	666 667	1 162 500		4 691 479	990 184			7 510 830
	Raised earth embankment	12 500 000	6 355 440	30 986 800	411 250	1 629 754	533 333	192 000	52 608 577
	Timber piling				1 570 133				1 570 133
Fluvial total		15 825 000	9 377 268	33 864 200	19 003 099	11 452 161	3 899 999	6 380 001	99 801 728
Other	Control structures: barriers/sluice/control					1 214 818			1 214 818
	Control structures: sluice/control/dropboards				47 667				47 667
	Control structures: weirs/sluice/control							5 333 333	5 333 333
	Debris, blockage clearance etc				0				0
	Flood defence gates		25 000	19 500			16 667	254 167	315 334
	Flood storage & alleviation scheme (no)				342 500				342 500
	Outfalls	2 386 250	30 730	370 667	86 892		1 062 500	187 500	4 124 539
	Pumping stations	200 000	855 000	60 000	700 000	715 200	410 000	69 444	3 009 644
	Reservoirs						1 000 000		1 000 000
	Sluices	130 000	109 200				750 000		989 200
	Trash screens		134 810					8400	143 210
	Weirs		3 843 000	164 700				1 650 000	5 657 700
Other total		2 716 250	4 997 740	614 867	1 177 059	1 930 018	3 239 167	7 502 844	22 177 945
Sea	Beach	239 760			0	931 714			1 171 474
	Beach (shingle)						1 666 667		1 666 667
	Beach (shingle) and control structure						2 475 000		2 475 000
	Beach + control structure	2 809 319				261 333			3 070 652
	Counterwalls						500 000		500 000
	Dunes	0		0	0				0
	Grassed/clay embankment	1 122 735	221 250	39 667	0				1 383 652
	High ground		0						0
	Offshore breakwaters						50 000		50 000
	Protected embankment	3 661 611	63 800	396 000	198 333	263 015	750 000		5 332 758
	Saltmarsh						166 667		166 667
	Seawall	8 583 575	192 500	14 020	406 500	1 472 746	3 000 000		13 669 341
Sea total		16 416 999	477 550	449 687	604 833	2 928 807	8 608 334	0	29 486 210
Tidal	Composite structures							8580	8580
	Ditch			170 000					170 000
	Raised earth embankment	5 473 333	2 443 000	2 586 267	143 063	801 168	4 166 667	5 180 000	20 793 498
	Revetted embankment	2 783 333	2 361 000	540 000	204 795		3 355 000	6 308 000	15 552 128
	Seawall	826 667	534 000	217 310	443 484	788 865		14 466 667	17 276 993
	Steel sheet piling	1 791 667	137 200	694 808	428 750		1 430 000	8 207 500	12 689 925
	Timber piling	100 000		367 500	13 653			283 333	764 486
Tide total		10 975 000	5 475 200	4 575 885	1 233 745	1 590 033	8 951 667	34 454 080	67 255 610
Grand total		45 933 249	20 327 758	39 504 638	22 018 736	17 901 019	24 699 167	48 336 925	218 721 493

A2 SUMMARY SHEETS OF SECONDARY MATERIALS

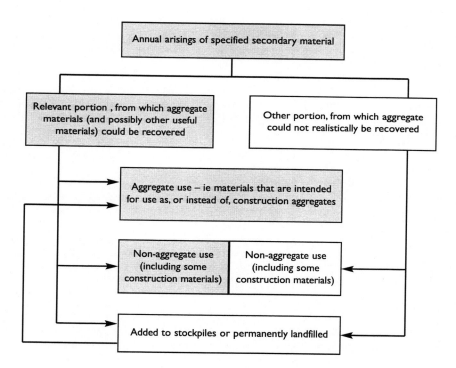

Figure A2.1   *Arisings flow chart*

In all cases, mt = million tonnes and kt = thousand tonnes.

Table A2.1   *Summary of attributes of china clay sand and rock*

Material type	CHINA CLAY SAND	
Relevant material and location	Rock and sand from china clay quarries in the South West.	
Tonnages in England and Wales in 2001	Arisings	c 22.6 mt
	Relevant portion	c 20.0 mt (89 per cent of arisings)
	Aggregate use	c 2.2 mt (11 per cent of relevant portion)
	Non-aggregate use	Nil
	Existing usable stockpiles	c 45–100 mt
Explanatory comments	Most china clay sand and rock is classified as a waste only because there is insufficient demand for it within its economic area of distribution. For every tonne of "good" china clay, roughly 9 t of by-product are generated (typically 4 t of coarse china clay sand, 4 t of coarse rock material (stent), and 1 t of micaceous residue).	
Arisings	**Located in South West** China clay and associated waste production vary from year to year, though the long-term trend is relatively stable. There may be a reduction in clay extraction tonnages from Cornwall consequent on an expansion of operations in Brazil.	
Stockpiles	**Located in South West** Annual arisings of china clay sand more than satisfy current limits of demand. The total volume of stockpiles is estimated at 600 mt, of which 20–60 per cent is usable, and some (not estimated) now lies under areas of nature conservation interest. No detailed survey has been undertaken of the extent of accessible and higher-quality stockpiles, but (unlike several other secondary materials) such stockpiles do exist. They may possibly amount to 45–100 mt.	
Use	**Predominantly in South West** China clay waste has an established sand and aggregate use. The rock from china clay waste can be screened to produce graded aggregate and sand. The coarse china clay sand is suitable for use as fine aggregate, or manufacture of concrete or concrete blocks. The remaining sand and gravel is suitable for construction and bulk fill. There is currently no use for micaceous residue (c 11 per cent of arisings), which, after settlement in lagoons, is disposed of.	
Trends in aggregate use	From 1999, utilisation increased by c 52 per cent to c 0.78 mt. Virtually all the material sold as aggregate was used locally in the SW region because of high transportation costs. Aggregate use is likely to increase significantly for the following reasons. 1 China clay sand is exempt from the £1.60/t Aggregates Levy (implemented April 2002), which increases the commercial feasibility of transporting the material outside the SW. 2 Investment in the development of the Port of Par, Cornwall, is aimed at increasing the export of sand and aggregates, although these plans are currently on hold. The proposal involves the construction of new railway and sea wharves, storage, conveyance and ship-loading facilities. Exports grew from 0 t in 1998 to c 200 000 t in 2003. Sales are expected to increase by c 80 per cent in 2004/5, with c 30 per cent of total sales being exported by sea.	
Future potential	To date, the main constraint on utilisation has been geography (cost of transport). With exemption from the Aggregates Levy and possible future investment in the Port of Par, china clay sand is becoming an increasingly competitive source of sand and aggregate. The feasibility of moving substantial quantities of material by rail from Cornwall to bulk fill projects in the south-east and south-west of England is being investigated. In April 2004 consideration was being given to the use of china clay sand for a major beach recharge project in Carlyon Bay, Cornwall.	
Data sources and assumptions used in previous rows	Consolidated figures and information from Imerys, Goonvean, and Watts, Blake, Bearne, plus information from *ENDS* and *Recycling*. BS 6543:1985	

Table A2.2 *Summary of attributes of slate aggregate*

Material type	SLATE AGGREGATE	
Relevant material and location	Waste material from slate mines and quarries in the North West, South West and North Wales.	
Tonnages in England and Wales in 2001	Arisings	c 6.3 mt
	Relevant portion	c 6.3 mt (100 per cent of arisings)
	Aggregate use	c 0.6 mt (9 per cent of relevant portion)
	Non-aggregate use	Nil
	Existing usable stockpiles	c 456 mt
Explanatory comments	None.	
Arisings	**North West, South West, North Wales** Arisings (94 per cent of "good" slate production) appear relatively stable. The English slate industry is entirely quarry-based, and small in comparison with its Welsh counterpart. About 85 per cent of English waste arisings are in Cornwall. All major quarries in North Wales have planning consents for 20+ years' extraction and scope for many more years of operation. Arisings here are expected to continue at above 4 mt/a.	
Stockpiles	**North West, South West, North Wales** Usable stockpiles are as follows: unspecified in North West (Cumbria); c 0.5 mt in South West (Cornwall); c 277 mt in North Wales (estimated by Arup as being recoverable without undue environmental effect). Relatively hard-to-use long-term stockpiles are as follows: c 5 mt in Cumbria; c 12 mt in Cornwall; c 453 mt in North Wales.	
Use	**Nationwide (mainly in areas of production – see above)** Slate waste has an established aggregate use in low-value applications. It complies with, and is accepted for, many applications within the *Specification for highway works* including sub-base, pipe beddings, drain stones, aggregate for coated macadam road-base, and aggregate for concrete products. The material is also used as low-grade general fill. Non-aggregate uses include mulch. Some quarries retain waste for site remediation under planning conditions.	
Trends in aggregate use	Although increasing, utilisation is still low because of the low quality and low value of slate aggregate, and its distance from main markets. For these reasons, aggregate use has occurred over the past 20 years only on a very local basis (20-mile radius). With exemption from the Aggregates Levy, transporting the material is becoming commercially feasible. Alfred McAlpine (the main producer of slate aggregate in North Wales) anticipates local road-delivered secondary aggregates will increase by 40–50 per cent. The company also plans to invest in railway infrastructure and to transport the material to north-west England, the Midlands and possibly the South East.	
Future potential	With exemption from the Aggregates Levy, and potential investment in railway infrastructure, aggregate use should rise, although it may be several years before significant increases occur. Developments such as thermal processing in rotary kilns to produce lightweight aggregate also increase the scope for further aggregate utilisation.	
Data sources and assumptions used in previous rows	Data from operators in Cumbria (Burlington and Kirkstone), Gwynedd (Alfred McAlpine) and from Gwynedd, Cornwall and Cumbria County Councils. Other information from *North Wales slate tips – a sustainable source of secondary aggregates* (Ove Arup, 2001) and *ENDS*. BS 6543:1985	

Table A2.3   *Summary of attributes of colliery spoil*

Material type	COLLIERY SPOIL	
Relevant material and location	Colliery spoil from deep coal mines in North East, Yorkshire and the Humber, West Midlands, East Midlands and South Wales.	
Tonnages in England and Wales in 2001	*Arisings*	c 7.5 mt
	Relevant portion	c 7.5 mt (100 per cent of arisings)
	Aggregate use	c 0.8 mt (11 per cent of relevant portion)
	Non-aggregate use	Nil
	Existing usable stockpiles	c 10–20 mt
Explanatory comments	Figures relate to colliery spoil (minestone) from deep mines. Opencast mines are generally required by the terms of their planning permissions to use all waste material in site restoration.	
Arisings	*Located in North East, Yorkshire and the Humber, West Midlands, East Midlands and South Wales* Arisings of minestone are determined by geology and are not in any consistent proportion to coal output.	
Stockpiles	*Located at active collieries (more widely for hard-to-use long-term stockpiles)* Usable stockpiles amount to c 20 mt (roughly 1 mt per working mine) and are found only at working mines (see above for geographical distribution). Most collieries intend to use stockpiled waste as bulk fill in site restoration. In terms of relatively hard-to-use long-term stockpiles, International Mining Consultants (ex-British Coal) consider that both "black" and "green" tips (located all over traditional mining areas) are at very much the same level as they were in 1990 (2000 mt and 1600 mt respectively in England and Wales. More precise estimates would require substantial research). Utilisation is likely only if a road or similar development requires a tip to be disturbed.	
Use	*Used at or close to active collieries (see above for locations)* Most colliery spoil is used as bulk fill, notably for constructing lagoons at collieries themselves. It is also used in landfill engineering (liner and capping systems), flood protection works, beach replenishment, and road construction (eg the Selby Bypass). In the near future it is possible that c 1 mt of minestone will be used in improvement works along the A1(M) between Redhouse and Dishforth.	
Trends in aggregate use	English deep-mined coal output fell from 19.3 mt in 1999 to 16.0 mt in 2001. This trend is unlikely to be reversed, and it is not thought that any new deep mines will be developed in the foreseeable future. Use of minestone has declined in recent years, though not quite so steeply as coal production itself. There are strong planning controls on lorry movements at some mines, which restrict utilisation. Minestone is virtually given away at present, making it uneconomic to dig up old spoil heaps, even if planning controls and minerals planning authorities would allow this.	
Future potential	Mining (and the associated arisings of colliery spoil) is likely to continue to decline. Employing colliery spoil in horticultural applications, and as a clay substitute, is in its infancy, but has the potential to make use of a substantial quantity of material. Various kiln-based processes are available to convert fines into a lightweight aggregate. Minestone can also be used as a feedstock for cement. The degree to which these potential uses are realised is dependent on economics. The greatest hindrance is the heat energy required to reduce water content. Planning restrictions make significant drawdown of even the most available stockpiles unlikely, unless they coincide directly with development. However, exemption from the Aggregates Levy has improved the commercial feasibility of using minestone as aggregate. It is the view of the industry that any moves towards performance specifications rather than "recipe" specifications in construction would also assist.	
Data sources and assumptions used in previous rows	Data from UK Coal Mining Ltd (12 mines), Coalpower (one mine), Betws Anthracite Ltd (one mine). Output for Goitre Tower Anthracite has been estimated. Geographical breakdown of usable stockpiles is based on an estimate of approximately 1 mt per active colliery.	

Table A2.4   *Summary of attributes of used tyres*

Material type	USED TYRES	
Relevant material and location	Scrap tyres arising nationwide.	
Tonnages in England and Wales in 2001	Arisings	c 400 kt
	Relevant portion	c 400 kt (100 per cent of arisings)
	Aggregate use	c 90 kt (22.5 per cent of relevant portion)
	Non-aggregate use	c 170 kt
	Existing usable stockpiles	c 14 million tyres in England and Wales
Explanatory comments	Figure for arisings excludes reused tyres. Used tyres are processed into rubber crumb before being used as aggregate.	
Arisings	**Nationwide** Arisings are growing at 2 per cent a year. The ban on landfilling of tyres (see below) will increase those available for recovery. Although this could lead to an increase in illegal dumping, moves towards producer responsibility may help to alleviate this problem.	
Stockpiles	**Significant stockpiles in Yorkshire and the Humber, East of England, South Wales** There are several large tyre dumps across England and Wales estimated to contain 14 million tyres: South Wales c 81 kt (9 million tyres) Yorkshire and the Humber c 18 kt (2 million tyres) East of England c 7 kt (0.8 million tyres). An additional six stockpiles in England and Wales hold more than c 0.90 kt (100 000 tyres). Intense fires deep inside some tyre dumps make recovery almost impossible. Widespread observations confirm that significant numbers of used tyres are used to hold down plastic sheeting on silage clamps on cattle farms.	
Use	**Nationwide** Some tyre crumb is used as a substitute for aggregate in the surfacing of sports and play areas, and in road surfacing. Some whole tyres are used in landfill engineering (in the drainage layer). The non-aggregate use figure given above (43.5 per cent) is broken down as 20 per cent fuel for cement kilns, 10.5 per cent recycled as retreads, 8 per cent incinerated with energy recovery, 2 per cent used in carpet underlay, 2 per cent exported, and 1 per cent used in pyrolysis.	
Trends in aggregate use	Road surfacing use is still at the trial stage in the UK, but is well established in France. The resultant surface is quieter and less reflective than conventional asphalt, but 10 per cent more expensive.	
Future potential	Potential for growth in aggregate substitution exists, but total amounts are likely to remain small. Interest in increasing recycling has been stimulated by the Landfill Directive (which calls for a ban on the landfilling of whole tyres by July 2003 and of shredded tyres by July 2006, though some sites will not have to comply until July 2007). The EU Directive on End of Life Vehicles also specifies targets for increasing reuse and recovery within this waste stream. Tonnages used in cement kilns are predicted to increase. Although initial investment in equipment is high, once installed, cement manufacturers can charge a gate fee for tyres and use them to replace coal or coke. The use of tyres in pyrolysis is also forecast to increase (up to 90 kt by 2003). Both of these uses will compete with aggregate use. The Environment Agency and DTI are looking at ways to encourage the use of re-treads. A £30 000 project has been launched to examine the use of tyres in river and coastal defences.	
Data sources and assumptions used in previous rows	Figures based on "medium recovery" forecasts from the Used Tyre Working Group July 2001 report, (<http://www.tyredisposal.co.uk>), which stresses the approximate nature of most data reported. Other information from RMC, ENDS, IWM, *Tyre trade news*. Estimates of relatively hard-to-use stockpiles from Environmental Data Interactive Exchange (EDIE) (tonnages based on the average tyre weighing 9 kg). Estimated geographical breakdown has been calculated by dividing total figures by population and should be treated with caution. *The pyrolysis of scrap automotive tyres* (Williams et al, 1990). *Fuel* 69:1474–1482.	

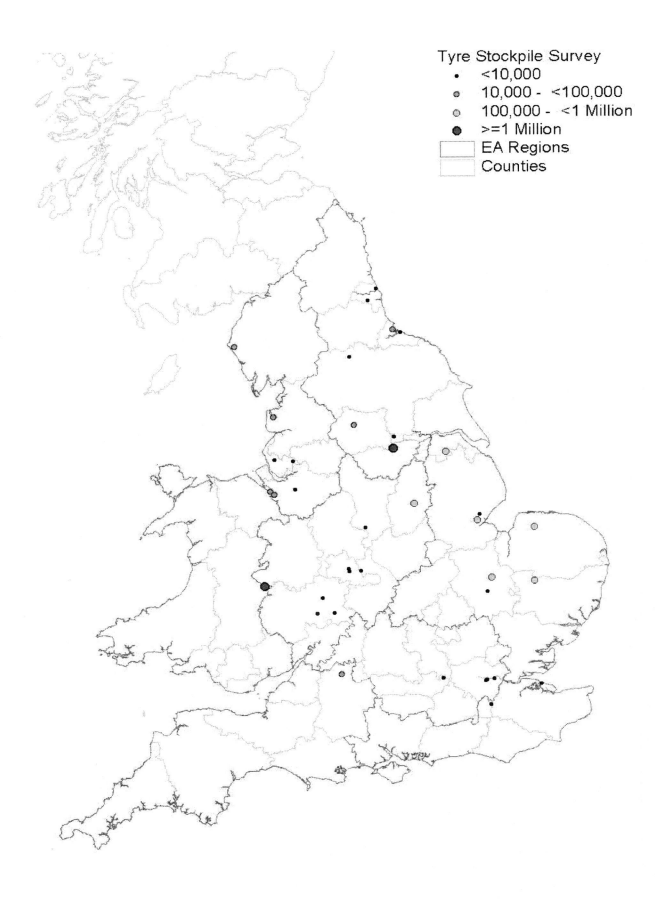

Figure A2.2   *Tyre stockpiles mapped on to Environment Agency and county boundaries*

Information taken from Tyres Stockpile Survey carried out in 2002 by AEA Technology for the Environment Agency.

CIRIA C590

Table A2.5   *Summary of attributes of glass cullet*

Material type	GLASS CULLET	
Relevant material and location	Glass from the municipal solid waste stream (MSW) in England and Wales.	
Tonnages in England and Wales in 2001	Arisings	c 2.2 mt
	Relevant portion	c 2.2 mt (100 per cent of arisings)
	Aggregate use	c 85 kt (4 per cent of relevant portion)
	Non-aggregate use	c 0.6 mt
	Existing usable stockpiles	c 20–30 kt
Explanatory comments	Glass for recycling is segregated at source, separated in municipal reprocessing facilities, or collected from bottle banks. The estimate of 2.2 mt is post-consumer container glass in the MSW stream. This could be an underestimate as it excludes illegal imports.	
Arisings	**Nationwide.** Arisings amount to a few hundred tonnes per plant (material reprocessing facilities (MRFs)/collection sites) per month. Fluctuations in arisings are dependent on packaging trends, consumer demand (dictated by factors such as public holidays, particularly around Christmas), and on public participation in the segregation of glass at source.	
Stockpiles	**Nationwide (but negligible).** Usable stockpiles are very small (several tens of thousands of tonnes across England and Wales). Organisations processing glass build up stocks before they can market products. As a result there is a considerable time lag between arising and use of recycled glass throughout the industry. Around 1.3 mt of glass is landfilled each year in mixed waste, but could not be recovered.	
Use	**Nationwide.** There is a general consensus that it is preferable to recycle container glass back into new containers, rather than reusing it for other purposes. Around 0.6 mt of glass was recycled back into the container industry in 2001. Some 40 kt was used in glass-fibre manufacture and c13 kt was used in other markets such as water filtration. Glass container manufacturers in the UK mainly require recycled flint (clear) and amber (brown) glass, but around 70 per cent of all collected glass is green. Much of the 0.2 mt of coloured glass arising from the commercial/industrial sector is unsegregated. The glass aggregate industry argues that green and mixed glass can be used for aggregate without competing with the container industry, although it may compete with the glass-fibre industry. Around 85 kt of glass was used as aggregate in 2001, and although increasing, it is constrained by a lack of supply. Recycled glass can be used as a substitute for aggregate in asphalt production to replace up to 30 per cent of the aggregate. Other established aggregate uses include sharp sand (to BS 7533-1 to 11:1997–2004) for bedding beneath block paving and pipework, abrasive uses (such as shot-blasting), non-structural pre-cast concrete (up to 20 per cent of sand) and landscaping products (chippings, paving, tiles and similar).	
Trends in aggregate use	The amount of glass used as aggregate has rapidly increased over the last few years, and WRAP estimates that it has the potential to reach several mt. Most potential uses are listed above (under "Use").	
Future potential	The current constraint is that demand outstrips supply. Increasing the use of recycled glass as aggregate is dependent on a rise in collection rates to meet demand. Other barriers to growth are listed below. Many standards are material-specific and will need to be amended to include glass. There are no accepted standards for glass in terms of its physical and chemical properties. Some recycled glass (particularly > 5 mm) has a plate-like shape with a high flakiness index, which limits the applications. The silica in glass can react with the alkali in cement, causing cracks and unacceptable damage to concrete. This alkali-silica reaction (ASR) does not appear to occur with fine-sized glass (< 2–3 mm), and there is some evidence that additions of very fine glass (< 600 µm) can actually increase concrete strength. The use of glass as concrete aggregate (see above) is being developed by Columbia University (USA), Sheffield University and Dundee University. Developing aggregate uses for recycled glass include drainage aggregate, filtration media and insulation. There are many additional potential uses for recycled glass, some of which are still in the development phase while others have not yet been proven commercially in the UK. These include applications such as a fluxing agent in the manufacture of bricks and ceramics; a filler for paints and plastics; a raw material for the production of zeolites (used in detergents); and a rooting medium for use in horticulture. The London Remade initiative (to increase collection of glass in London's commercial/industrial sector; <http://www.londonremade.com>) is expected to divert 40 000 t of additional waste glass from London boroughs into construction over the next two years.	
Data sources and assumptions used in previous rows	Information from: British Glass, Valpack, WRAP, RMC, Day Group, Windmill Aggregates, *ENDS report*, *Construction news*, Columbia University website <http://www.columbia.edu/>. Estimated geographical breakdown (see below) has been calculated by dividing total figures by population, and should be treated with caution. BS 6543:1985	

Table A2.6   *Summary of attributes of fired ceramic waste*

Material type	CERAMIC WASTE	
Relevant material and location	Ceramic waste from the brick, tableware and sanitary industries nationwide, with local concentrations in the West Midlands and East of England.	
Tonnages in England and Wales in 2001	*Arisings*	c 100 kt
	Relevant portion	c 100 kt (100 per cent of arisings)
	Aggregate use	c 90–100 kt (90–100 per cent of relevant portion)
	Non-aggregate use	Nil
	Existing usable stockpiles	Working stockpiles only. No reliable quantitative estimates exist
Explanatory comments	None.	
Arisings	**Nationwide (local concentrations in West Midlands and East of England)** Waste as a percentage of saleable output continues to decline as a result of efficiency gains and increasing utilisation within the ceramics sector itself. Saleable output depends heavily on house-building and similar. Brick-making represents the largest part of the above figures (with fired waste estimated at <1 per cent of 7.2 mt of bricks).	
Stockpiles	**Nationwide (local concentrations in West Midlands and East of England)** Some ceramic waste is held in dedicated landfills in the Potteries (eg Shraff Tip). No easily accessible stockpiles are held.	
Use	**Nationwide** The main use is as bulk fill for roads and paths. Fired brick waste is crushed and used as aggregate, including a significant proportion in brick-pit and landfill haul roads. Fines from waste Fletton brick can be used as a cement replacement. Tableware waste (a few kt) is used in tile manufacture, sub-base material in footpaths and in surfacing landfill haul roads. All fired waste from clay pipe manufacture is ground down into "grog" (sand aggregate substitute), which is used as filler in the manufacture of new clay pipes. Similarly, all tile waste is incorporated back into the production process.	
Trends in aggregate use	Utilisation is already high, with little scope for it being raised.	
Future potential	Further decline in arisings likely.	
Data sources and assumptions used in previous rows	Information and statistics from the British Ceramic Confederation and tableware, pipe, tile and brick manufacturers. It has been assumed that tableware/tile waste used in the manufacture of new tiles constitutes aggregate use. Clay pipe waste going directly back into the production process (before leaving the factory gates) has not been included in the estimate of aggregate use.	

Material type	SPENT FOUNDRY SAND	
Relevant material and location	Spent foundry sand from the castings industry nationwide (mainly in the Midlands and North).	
Tonnages in England and Wales in 2001	Arisings	c 0.9 mt
	Relevant portion	c 0.9 mt (100 per cent of arisings)
	Aggregate use	c 0.09–0.18 mt (10–20 per cent of relevant portion)
	Non-aggregate use	Nil
	Existing usable stockpiles	No reliable quantitative estimates exist
Explanatory comments	None.	
Arisings	**Nationwide (mainly North West, North East, Yorkshire and the Humber, West Midlands, East Midlands)** The castings industry is declining as a result of the strong pound and changes in the car industry (which is using more aluminium and plastics, and less steel). Sand reuse has increased in response to the landfill directive, which has stimulated investments in thermal recycling processes. The castings industry is likely to decline further due to the climate change levy. However, the climate change levy also discourages further thermal recycling processes (which can push in-foundry reuse rates from 50 per cent to 90 per cent), so the decline is likely to be in direct proportion to the economic health of the castings sector. The split between England and Wales is unknown, but is estimated to be within the order of 98:2, based on incomplete returns from selected suppliers of new foundry sand.	
Stockpiles	**Nationwide** Many small, geographically scattered usable stockpiles exist at castings works. No quantitative estimates exist. Use of these small stockpiles outside of the industry is inhibited by high transport costs, making landfill of small stockpiles an easier means of dealing with the waste. The majority of reclaimed spent foundry sand is therefore from large foundries. Anecdotal evidence of relatively hard-to-use long-term stockpiles suggests that there are some substantial dedicated landfills of spent foundry sand at major castings works, possibly amounting to several mt. No consolidated/detailed estimates of availability and quality have been found.	
Use	**Nationwide** Foundry sand only equates to a very small percentage of sand used by the construction industry. The main uses of spent foundry sand are in block manufacture and ready-mix concrete. A smaller quantity is used in asphalt manufacture.	
Trends in aggregate use	The uses of foundry sand are now relatively well established. The Landfill Tax has stimulated lower-value applications, as will exemption from the Aggregates Levy.	
Future potential	Further decline in arisings is likely. The use of spent foundry sand in land remediation projects by mixing with organic waste has been investigated, as has mixing it with bentonite for use as a landfill liner	
Data sources and assumptions used in previous rows	Information and statistics from the Silica and Moulding Sands Association, the *Foundry trade journal*, WBB Minerals and Mansfield Sand.	

Table A2.8   *Summary of attributes of blast-furnace slag*

Material type	BLAST-FURNACE (BF) SLAG		
Relevant material and location	Slag from integrated steelworks with blast furnaces (for iron processing prior to steel-making) in Yorkshire and the Humber and South Wales.		
Tonnages in England and Wales in 2001	Arisings	c 3.0 mt	
	Relevant portion	c 3.0 mt (100 per cent of arisings)	
	Aggregate use	c 0.9–1.2 mt (30–40 per cent of relevant portion)	
	Non-aggregate use	c 1.8–2.1 mt (60–70 per cent, by difference)	
	Existing usable stockpiles	No reliable quantitative estimates	
Explanatory comments	Blast-furnace (BF) slag is either air-cooled or quenched: the choice is the operator's and dictates the end use of the material. Quenched slag is then ground up and used as ground granulated blast-furnace slag (GGBS), a cement replacement (ie not an aggregate). Air-cooled slag is primarily used as roadstone aggregate.		
Arisings	***Located in Yorkshire and the Humber and South Wales*** In 2001 Corus operated two integrated plants in Yorkshire and the Humber (Teesside and Scunthorpe) and two in South Wales (Llanwern and Port Talbot). Llanwern ceased production in July 2001. Production at the Port Talbot plant is due to increase with the recent refurbishment of Port Talbot No 5 blast furnace, which was closed for part of 2001 after an explosion. The amount of slag produced per tonne of iron is typically 295 kg, and iron production in England and Wales was 9.9 mt in 2001. The split between England and Wales for iron production was 67:33. The explosion at Port Talbot No 5 blast furnace, together with the closure of production at Llanwern in 2001, caused a significant decrease in Welsh production. When Port Talbot is fully operational the split should return to around 67:33 unless there are major changes in English production.		
Stockpiles	***Located in Yorkshire and the Humber and South Wales*** There are some usable stocks of air-cooled slag at steelworks. No reliable quantitative estimates for such stocks are available. There are also some long-term relatively hard-to-use stockpiles of air-cooled slag at steelworks. No reliable quantitative estimates for such stocks are available.		
Use	***Nationwide.*** Air-cooled slag can be used as roadstone aggregate. A significant part of the arisings of BF slag in Wales (estimated to be approximately 1.58 mt), is supplied as aggregate and GGBS to customers in England.		
Trends in aggregate use	A decline in slag arisings and strong growth in the GGBS market has reduced the amount of BF slag available for use as aggregate. During the 1990s, BOF steel production was relatively stable, and averaged around 13 mt a year in England and Wales. By 2001 it had fallen to 10.4 mt. The trend for BF iron was directly comparable. Recycling as GGBS is understood to have been rising relative to aggregate production. Details are commercially sensitive, but the share of BF slag used as GGBS in England and Wales is believed to be approximately 60–70 per cent. The share used as aggregate is therefore estimated at 30-40 per cent.		
Future potential	Further decline is likely as a result of steel industry trends.		
Data sources and assumptions used in previous rows	Information from Tarmac (which takes all of Corus's air-cooled slag), Corus, and the Iron and Steel Statistics Bureau's website <http://www.issb.co.uk>. BS 6543:1985.		

Table A2.9   *Summary of attributes of basic oxygen furnace bottom slag*

Material type	BASIC OXYGEN FURNACE (BOF) STEEL SLAG	
Relevant material and location	Steel slag from integrated steelworks (which contain both blast furnaces for iron processing and basic oxygen furnaces for steel-making) in Yorkshire and the Humber and South Wales.	
Tonnages in England and Wales in 2001	Arisings	c 1.0 mt
	Relevant portion	c 1.0 mt (100 per cent of arisings)
	Aggregate use	c 0.98 mt (98 per cent of relevant portion)
	Non-aggregate use	c 0.02 mt
	Existing usable stockpiles	No reliable quantitative estimates
Explanatory comments	All (or virtually all) available BOF steel slag is understood to be processed. It is generally left to undergo natural weathering before and after processing	
Arisings	**Located in Yorkshire and the Humber and South Wales** In 2001 Corus operated two integrated plants in Yorkshire and the Humber (Teesside and Scunthorpe) and two in South Wales (Llanwern and Port Talbot). Llanwern ceased production in July 2001. Production at the Port Talbot plant is due to increase with the recent refurbishment of Port Talbot No 5 blast furnace, which was closed for part of 2001 after an explosion. In 2001 steel production in England and Wales was 13.5 mt, of which 10.4 mt was BOF steel, and the amount of slag produced per tonne of basic oxygen furnace (BOF) steel is estimated to be roughly 150 kg. About 10 per cent of BOF steel slag is reused in the production process and has been excluded from arisings. The split between England and Wales for steel production was c 67:33 in 2001. The explosion at Port Talbot No 5 blast furnace, together with the closure of production at Llanwern in 2001, caused a significant decrease in Welsh production. When Port Talbot is fully operational the split should return to around 67:33 unless there are major changes in English production.	
Stockpiles	**Located in Yorkshire and the Humber and South Wales** There are some usable stocks of air-cooled slag at steelworks (separate from those left for a limited period of time to "weather" before and after processing). No reliable quantitative estimates for such stocks are available. There are also some relatively hard-to-use long-term stockpiles of air-cooled slag at steelworks (see above for geographical distribution). No reliable quantitative estimates for such stocks are available.	
Use	**Nationwide** Around 98 per cent of BOF steel slag is air-cooled and used as roadstone aggregate. The remaining 2 per cent is used in agriculture (non-aggregate use). A significant portion of arisings in Wales (estimated to be approximately 90 per cent) is supplied as aggregate to customers in England.	
Trends in aggregate use	During the 1990s, BOF steel production was relatively stable, and averaged around 13 mt a year in England and Wales. By 2001 it had fallen to 10.4 mt. BOF steel slag is a high-quality material and has a well-established aggregate use. However, demand outstrips supply, and the quantities used as aggregate are governed by arisings.	
Future potential	Further decline is likely as a result of steel industry trends.	
Data sources and assumptions used in previous rows	Information from Tarmac (which takes all of Corus's air-cooled slag), Corus, and the Iron and Steel Statistics Bureau's website <http://www.issb.co.uk>. BS 6543:1985.	

Material type	ELECTRIC ARC FURNACE (EAF) STEEL SLAG	
Relevant material and location	Slag from electric arc furnace (EAF) steel plants in Yorkshire and the Humber and the South East.	
Tonnages in England and Wales in 2001	Arisings	c 0.28 mt
	Relevant portion	c 0.28 mt (100 per cent of arisings)
	Aggregate use	c 0.28 (100 per cent of relevant portion)
	Non-aggregate use	Nil
	Existing usable stockpiles	Minimal. No reliable quantitative estimates
Explanatory comments	None.	
Arisings	**Located in Yorkshire and the Humber and South East** EAF steel production in England and Wales in 2001 is estimated to have been c 3.1 mt. The ratio between EAF steel production and slag is roughly 10:1. EAF steel slag has a well-established aggregate use. All (or virtually all) available slag is processed.	
Stockpiles	**Located in Yorkshire and the Humber, South East and South Wales** Small short-term usable stockpiles exist at operational steel plants, and a small amount is stockpiled at a recently closed plant in South Wales. No reliable quantitative estimates for these stocks are available. No relatively hard-to-use long-term stockpiles are thought to exist.	
Use	**Nationwide** EAF steel slag is a high-quality material used in asphalt mixes (particularly in the wearing course of roads) and for surface dressing where skid resistance is important.	
Trends in aggregate use	There has been a gradual decline in EAF steel production (and therefore slag arisings) in recent years. In England and Wales EAF steel production averaged 4.2 mt/a during the 1990s, dropping to 3.7 mt in 1999 and 3.1 mt in 2001. With the closure of the steel mill in Cardiff in 2001, EAF steel slag is no longer produced in Wales.	
Future potential	Further decline is likely as a result of steel industry trends.	
Data sources and assumptions used in previous rows	Information from Heckett MultiServ (Steelphalt) Ltd. BS 6543:1985.	

Table A2.11   *Summary of attributes of power station furnace bottom ash*

Material type	POWER STATION FURNACE BOTTOM ASH (FBA)	
Relevant material and location	Bottom ash from coal-fired power stations in the North West, North East, Yorkshire and the Humber, West Midlands, East Midlands, East of England, South East and South Wales.	
Tonnages in England and Wales in 2001	Arisings	c 0.98 mt
	Relevant portion	c 0.98 mt (100 per cent of arisings)
	Aggregate use	c 0.97 mt (99.3 per cent of relevant portion)
	Non-aggregate use	Nil
	Existing usable stockpiles:	No reliable quantitative estimates
Explanatory comments	None.	
Arisings	*Located in North West, North East, Yorkshire and the Humber, West Midlands, East Midlands, East of England, South East and South Wales* Coal burned in power stations (and consequently FBA production) declined by over 50 per cent between 1990 and 2000 (an annual decline of roughly 8 per cent). There was a slight increase in coal-fired electricity production during 2001.	
Stockpiles	*Located at coal-fired power stations (see above for locations, plus some former sites)* No hard-to-use long-term stockpiles exist, although small amounts of bottom ash may be found in fly ash landfills as a drainage material. Only 6.6 kt/a of FBA is thought to go to landfill.	
Use	*Nationwide* FBA utilisation has remained consistently very high. Virtually all FBA is used in lightweight block manufacture.	
Trends in aggregate use	Use of FBA is fairly stable.	
Future potential	A further decline in ash arisings is likely, dependent on market forces (eg cost of alternative fuels such as gas), and UK Government policy on electricity generation.	
Data sources and assumptions used in previous rows	Data from UK Quality Ash Association (UKQAA) and DTI website <http://www.dti.gov.uk>. Also see environmental reports from Innogy, TXU, AEP, LPC and PowerGen. Geographical breakdown of estimated arisings based on installed capacity of power stations at end of May 2001 and figures from company websites.	

Table A2.12   *Summary of attributes of pulverised fuel ash*

Material type	PULVERISED FUEL ASH (PFA)	
Relevant material and location	Fly ash from coal-fired power stations in the North West, North East, Yorkshire and the Humber, West Midlands, East Midlands, East of England, South East and South Wales.	
Tonnages in England and Wales in 2001	Arisings	c 4.87 mt
	Relevant portion	c 4.87 mt (100 per cent of arisings)
	Aggregate use	c 1.66 mt (34 per cent of relevant portion)
	Non-aggregate use	c 0.83 mt
	Existing usable stockpiles	c 55 mt
Explanatory comments	None.	
Arisings	**Located in North West, North East, Yorkshire and the Humber, West Midlands, East Midlands, East of England, South East and South Wales** Coal burned in power stations declined by more than 50 per cent between 1990 and 2000 (an annual decline of roughly 8 per cent). There was a slight increase in coal-fired electricity generation during 2001 and therefore PFA production. For PFA to be used in cementitious applications, base load stations (stations that operate continuously) are preferred. Stations that operate on an irregular basis produce a PFA more suitable for grouting and fill applications.	
Stockpiles	**Located at coal-fired power stations** Estimates of usable stockpiles amount to approximately 55 mt. These stockpiles are at coal-fired power stations. Many mt of fly ash could be accessed by stripping off topsoil. In terms of relatively hard-to-use long-term stockpiles, much fly ash has been "sterilised" (eg half of the 9 mt landfilled at Drax has been designated an SSSI).	
Use	**Nationwide** The most common use of PFA is in lightweight concrete block manufacture. This is followed by use as a cementitious material in ready-mixed and precast concrete (ie not as an aggregate), the grouting of mines and caverns, and finally as a fill material. PFA that has been landfilled can be recovered and used for grout or fill applications after screening. PFA is also used to make lightweight aggregate, commonly known as Lytag.	
Trends in aggregate use	The UKQAA is reporting a gradual increase in PFA utilisation in road sub-base construction. Non-aggregate use in cementitious applications is also increasing. Grouting and block manufacture usage remains stable. These trends have been offset by the difficulties in obtaining environmental approvals when used for unbound applications, eg as a fill material.	
Future potential	Further decline in ash arisings dependent on market forces, eg cost of alternative fuels such as gas, and Government policy on electricity generation. Recovering PFA from some old landfills is possible. This practice is carried out in France, and planning permission has recently been granted for Alcan Smelting and Power to recover c 3 mt of PFA from four waste sites in Northumberland for sale to the construction industry.	
Data sources and assumptions used in previous rows	Data from UK Quality Ash Association (UKQAA) <http://www.ukqaa.org.uk> and DTI website <http://www.dti.gov.uk>. Also see environmental reports from Innogy, TXU, AEP, LPC and PowerGen. Geographical breakdown of estimated arisings and stockpiles based on installed capacity of power stations at end of May 2001 and figures from some power companies' websites. BS 6543:1985.	

Table A2.13   *Summary of attributes of municipal solid waste incinerator bottom ash*

Material type	MUNICIPAL SOLID WASTE INCINERATOR BOTTOM ASH (IBA)	
Relevant material and location	Bottom ash from MSW incinerators in the North West, North East, Yorkshire and the Humber, West Midlands, East Midlands, East of England and London.	
Tonnages in England and Wales in 2001	Arisings	c 0.62 mt
	Relevant portion	c 0.62 mt (100 per cent of arisings)
	Aggregate use	c 0.38 mt (61.3 per cent of relevant portion)
	Non-aggregate use	Nil
	Existing usable stockpiles	Negligible. No reliable quantitative estimates
Explanatory comments	The estimates above refer to "new generation" MSW incinerators and exclude the manufacture of refuse-derived fuel.	
Arisings	**Located in North West, North East, Yorkshire and the Humber, West Midlands, East Midlands, East of England, London** Municipal waste is increasing and the EU Landfill Directive requires much more to be diverted from landfill over the next 20 years. The current capacity for municipal waste incineration is 2.7 mt/a, less than 10 per cent of the waste produced. The Environment Agency estimates that the amount of waste incinerated or recovered by other means may reach 10 mt/a by 2010. This could result in MSW incinerator bottom ash arisings of c 2.5 mt/a by 2010.	
Stockpiles	**Negligible** Working stockpiles are small and short-term. In August 2001, the Environment Agency announced that bottom ash is regarded as waste after it leaves processing plants until it is incorporated into construction products. Stockpiling of the material therefore requires a waste management licence. Prior to this there were reported to be instances of MSW incinerator fly and bottom ash being stockpiled, but such cases were rare. Although stockpiles may amount to a few thousand tonnes at some incinerator sites, quantities are otherwise small. Relatively hard-to-use long-term stockpiles do not exist and no unused bottom ash has been disposed of to dedicated landfills.	
Use	**Nationwide** The main uses for MSW incinerator bottom ash are road surfacing (asphalt layer) and concrete blocks. Other applications include bulk fill, road base material, and daily cover materials for landfill sites.	
Trends in aggregate use	Driven by factors such as increasing costs of disposal, and exemption from the Aggregates Levy, the use of MSW incinerator bottom ash as an aggregate is likely to increase. There have been several successful projects using IBA as aggregate that have been publicised by the Aggregates Information Service, including several instances of road construction.	
Future potential	Arisings and aggregate use are likely to increase, so long as regulatory constraints do not become too onerous. However, future potential is also affected by past history, and over recent years there has been some negative publicity concerning the use of MSW incinerator fly ash and bottom ash. Anxiety has centred on potential contamination with dioxins as a result of mixing fly and bottom ash. In 1998 concern was also expressed in the House of Commons about the use of ash as daily cover at a landfill, and the possibility that it had earlier been used to create site roads at a nearby site.	
Data sources and assumptions used in previous rows	Figures from the Environment Agency's returns, website, and report entitled *Solid residues from municipal waste incinerators in England and Wales* (Environment Agency, 2002). Other information from the Aggregates Information Service publications and ENDS. Estimate for arisings is based on EA data for 2000. Estimate for aggregate use is based on an average taken from three incinerators and two recycling facilities, so confidence in the estimate is not high. BS 6543:1985.	

A3　　　　　**CASE STUDIES**

Betteshanger Colliery, Kent – colliery spoil used as bulk fill in beaches for sea defence/flood prevention

Colliery spoil has been utilised as bulk fill (or substitute) for quite substantial flood and coastal defence structures. When the sea defences failed in front of the Betteshanger colliery, near Deal in Kent, during a winter storm and high tides in January 1978, about two million cubic metres of water flowed inland overtopping secondary defences, flooding structures and covering 300 ha of farmland. The Betteshanger tip (approximately 500 ha in area, 10 m high) was within 2 miles of the site and closer than any quarry or borrow site. This convenient resource was used as a large-volume beach core material, displacing the equivalent volume of shingle for redeployment on the seaward face. It also supported a new road of crushed rock laid slightly to landward on the top of the bank to provide access for machinery involved in replenishment and other maintenance schemes in the future. The whole scheme used 85 000 m³ of colliery spoil and 20 000 m³ of rip-rap.

(Hamilton, 1984)

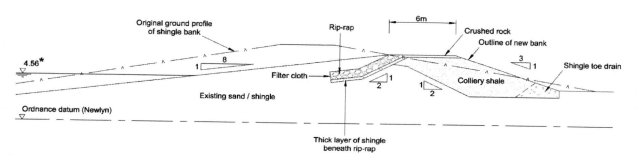

Figure A3.1   *Typical cross-section of the Betteshanger Sea Defence, Deal, Kent*

County Durham beaches – colliery spoil used (inadvertently) for beach replenishment

Colliery spoil tipped on to beaches at Dawdon and Easington, County Durham (obtained from Wearmouth discards), was analysed for physical breakdown by slake durability tests (McDermott, 1987). The shale was found to break down quite easily along fissile planes. Use of this material here has changed the morphodynamics of these beaches and now that tipping has stopped there is a strong likelihood that an equilibrium state will not be reached for a long time.

(Humphries *et al*, 1996)

Brighton Marina, Sussex – C&D waste recycling on site for sea defence (seawall)

The existing seawall, promenade and rear splashwall between Brighton Marina and Ovingdean on the south coast of England were some 70 years old. Necessary repairs after storm damage were becoming increasingly frequent and more extensive. A renovation of the defence system involved the encasement of the front face of the existing seawall, the removal and replacement of the bullnose, the raising of the promenade and the replacement of the rear splashwall.

The removal of demolished concrete from the site was likely to incur significant disposal costs, in the form of transport and Landfill Tax, if it could not be sold as fill for other works.

The anticipated effects of a rise in sea level rise now has to be taken into account when carrying out government grant-aided coastal defence schemes. To this end, it seemed appropriate to raise the crest level of the seawall and promenade as part of the improvement scheme. Raising the rear splashwall, although desirable, was not possible because of environmental constraints.

An assessment of concrete arisings from the demolition works conducted at the planning stage of this project enabled identification of potential reuse within the works in three ways.

1   Some of the larger sections of the demolished bullnose were suitable as secondary armour under the new rock armour revetment (replacing importation of granite armour from Norway).

Figure A3.3a   *Renovation of Brighton–Ovingdean sea defence wall – bullnose sections earmarked for secondary armour (courtesy Posford Haskoning)*

2   Some of the demolition arisings – concrete blocks and the friable no-fines concrete from behind the existing splashwall – were suitable as fill material within the two box-structure ramps leading from the promenade on to the foreshore. The fill is required to provide deadweight to stop the structures moving under wave loading.

3   Crushed concrete blocks were also used to raise the promenade by 600 mm, to take some account of sea level rise over the coming 50 years.

Figure A3.3b   *Foundations for the promenade ramps at Brighton (courtesy Posford Haskoning)*

Case Study A3.4

Minehead, Somerset – C&D waste use for sea defence (seawall)

Between 1997 and 1999, 1.8 km of new sea defences comprising a raised wall, rock groynes and renourished beaches were constructed at Minehead at a cost of £12 million.

Both the design and construction processes involved environmental considerations, one of which was the reuse of existing materials and minimising off-site disposal.

For construction of the new wall to take place, the existing seawall had to be demolished. The decision to recycle the old wall was driven primarily by economics. It was more than £50 000 cheaper to bring a crusher to the site and reuse the material as a sub-base in the new works than dispose of the material off site and import new aggregate. In total an estimated 10 000 t of material was recycled.

The process required some forward planning and co-operation between the client, designer and contractor. However, no difficulties were encountered and the operation was a success. The obvious advantages relating to sustainability, economics and public relations flowed from the decision to recycle.

Figure A3.4a   *Crushing concrete from the old seawall into aggregate for reuse*

Another major consideration was minimising the environmental impacts from transporting the 70 000 t of primary rock armour and core rock and the 30 000 t of additional aggregate required for the scheme. By road, this would have involved around 5000 return journeys, with obvious impacts on the environment. The Environment Agency took a proactive role and investigated the possibility of transporting the materials by rail. A new railway siding was constructed within the contractor's compound and the rock was sourced from two quarries in the Mendips that benefited from existing railheads. The materials were delivered by rail in 500 t units, which proved to be a reliable and cost-effective method with significant environmental benefits (Masters, 2001).

(Information courtesy of the Environment Agency.)

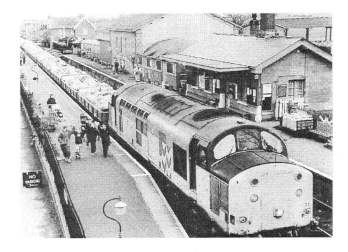

Figure A3.4b   *Delivery of primary rock armour by rail (courtesy of the Environment Agency)*

Figure A3.4c   *Minehead sea defences on completion of works (courtesy of the Environment Agency)*

Norfolk coast – C&D waste (concrete blocks) cliff toe protection (crib-work)

Figure A3.5   *Concrete blocks used to fill crib-work on the Norfolk coast*

C&D concrete blocks used to fill crib-work as a low-grade sea defence at the back of the beach. These are essentially cliff toe protection features but do experience a moderate wave climate on the coast of Norfolk.

Hayling Island, Hampshire – C&D waste used in seawalls for saline lagoons

In 1980 a private company approached Havant Borough Council with a view to restarting the shellfish farming industry at the Langstone Oyster Beds, Hayling Island, Hampshire. Construction began in 1981 when new material (derived from building rubble) was placed directly on top of the old, eroded shingle embankments. A total of 100 000 t of material was imported until in 1982 it was realised, belatedly, that the planning consent had erroneously quoted the permitted height as 5 m above Ordnance Datum. It had been intended that the banks be at a level approximating to high tide (5 m above Chart Datum). The banks were thus some 2.7 m higher than anticipated.

The operating company ceased trading in 1987, before it had completed finishing works such as placing protective shingle to the seaward face of the embankments and adding topsoil to landscape the site. This left the site overall in breach of the planning consent.

As landowner, the council was left with a dangerous legacy. In the intervening years, access by the public had become unrestricted for recreational use. However, the new material, which had merely been end-tipped from lorries, consisted of an apparently firm crust that in places overlay voids between large masonry blocks and was inherently unstable. Reinforcing rods and other metal waste within the rubble represented a hazard. Exposed to wind and wave action, the embankments eroded and quickly became dangerous to people walking upon them. The council tried to control the risk by erecting fencing and signs, but these were repeatedly vandalised. Subsequently, much time and money was spent before a solution was found and remediation began.

The oysterbeds lie within an area that is now designated:

- Site of Special Scientific Interest (SSSI)
- Special Protection Area (SPA)
- candidate Special Area of Conservation (cSAC)
- Ramsar wetlands site
- Area of Special Landscape Quality (ASLQ).

The harbour is home either permanently or on a migratory basis for tens of thousands of seabirds, attracted by a wealth of marine invertebrate life in the thousands of hectares of intertidal mudflats. It is also a popular area for both residents and visitors from farther afield due to the wildlife and landscape quality, which has been recognised by the borough council in its designation (ASLQ) under the local plan.

Figure A3.6   *Lagoon wall at Langstone Harbour oysterbeds, Hayling Island*

Case Study A3.7

Liverpool Bay – C&D waste and tin slag beach

In the early part of the 20th century no artificial defences existed to protect this part of the coastline, and the shoreline consisted of sand dunes that were eroded as the River Alt channel meandered along the frontage. As a result, several shoreline properties were lost to the sea. Slag from a local tin-smelting works and subsequently rubble were tipped on the shoreline in an attempt to arrest the erosion, but it was not until the Alt channel was diverted by the construction of the training bank to the north in the mid-1930s and linear defences were built across the frontage in 1960, that the integrity of the shoreline was safeguarded.

The beach lies within an area (s) classified under several designations:

- Site of Special Scientific Interest (SSSI)
- Coastal Area of International Importance (Ramsar)
- Special Area for Conservation (SAC)
- Special Protection Area (SPA).

Figure A3.7   *Construction and demolition rubble beach contaminated with tin slag, River Alt, Liverpool Bay*

Beesands, Devon – recovered rock (from site) reused in new revetment

The village of Beesands lies within Start Bay in South Devon, England. In the past, the village relied upon a wide beach for its coastal defence, but this slowly eroded away until the soft slope behind became vulnerable to erosion and overtopping. Rock had been tipped along the frontage periodically, generally as a reaction to emergencies. Following a partial failure of the defence during storms in the winter of 1989/90, a scheme was devised for a longer-term solution. This included a 500 m-long rock revetment backed over the central 260 m by a reinforced concrete wave-wall. The rock revetment contained approximately 20 000 t of filter material and 60 000 t of armour blocks. This comprised 28 000 m^3 of armour rock imported from Sweden, 1 500 m^3 of imported filter rock and 14 000 m^3 of rock recovered from the site.

Port Edgar, Lothian – floating tyre breakwater

A floating tyre breakwater was built in Port Edgar, Lothian, from truck tyres in 1979. The structure successfully reduced wave action in the harbour. Waves with periods of up to 5 seconds were damped significantly. The major problem encountered at this site was the growth of marine organisms on the structure causing a loss of buoyancy. This required regular maintenance.

(Motyka and Welsby, 1983).

Figure A3.9   *Floating tyre breakwater, Port Edgar, Lothian*

River Witham, Lincoln – tyre bales in flood embankment restoration scheme

Environmental concerns and structural design restrictions have led to the innovative use of tyre bales in a design solution for the restoration of 1.7 km of a flood defence barrier for Branston Island on the River Witham in Lincolnshire.

The embankment separates Branston Island, an emergency flood storage area, from the river. This embankment was badly eroded with a crest width of only 2.5 m. Works were required to increase this width to 4 m, reprofile the embankment, reinstate the berm and install toe protection. The embankment was sitting on a peat base, so to prevent slippage if using clay as the construction material, the embankment would have required reprofiling to a 1:4 slope ,enlarging the structural footprint area. This would have necessitated relocation of 11 kVA power lines and a soke dyke running parallel to the embankment, disrupting the local environment as well as increasing costs.

Figure A3.10   *Setting tyre bales into the embankment excavation (courtesy of the Environment Agency)*

By using tyre bales, a much steeper bank profile could be achieved, reducing the potential increase in footprint size and thus avoiding the need to relocate the soke dyke and power lines, so saving time and money. Less material was used, as tyres are much less dense than clay. The volume of material excavated from the bank to install the bales roughly equated to that required for capping the structure, so only an estimated 1000 m³ of supplementary soil excavated from a local borrow pit was required to provide adequate cover. The design of the bank allows for its surface to be seeded and for a degradable protective coir (coconut husk) matting to be laid to protect against soil erosion until vegetation becomes re-established. Work on this project had largely been completed by April 2004.

A programme of monitoring and recording is planned to study the physical performance of the bales and of any leachates appearing in drainage water or in the river.

Poole beaches, Dorset – maintenance dredgings used for beach replenishment

In the Port of Poole capital works in the approach channel to the harbour in 1988/89 resulted in 604 000 m³ out of 675 000 m³ of sand dredged being pumped on to Bournemouth beaches. Further capital works the following winter resulted in an additional 420 000 m³ being pumped ashore, due entirely to the success of the previous development. In 1991/92 developments to the navigation channel inside Poole Harbour resulted in 40 000 m³ of fine sand being jetted on to the beach at Sandbanks for Poole Borough Council (Humphreys *et al*, 1996).

Figure A3.11 shows a beneficial recharge scheme at Parkstone Yacht Club in the harbour. As a condition of gaining planning permission, the developers were obliged to create an artificial mudflat similar in area to the intertidal area lost through dredging for the marina.

Figure A3.11   *Mudflat creation scheme at Parkstone Yacht Club, Poole Harbour, Dorset*

Melton Mowbray, Leicestershire – C&D waste used as bulk fill in railway embankment fill in a flood alleviation scheme

The Melton Mowbray FAS was a scheme to defend the town from flooding by means of on-line flood storage. In addition to the earth embankment dam works, a railway line that crossed the footprint of the reservoir required stabilisation. The railway embankment shoulders were reinforced by filling with a free-draining material, thereby increasing its resistance to the effects of rapid drawdown and washout of fine material.

The material used came from the haul roads used to gain site access. These haul roads were constructed with a limestone base and granite scalpin wearing course. The granite scalpins were a by-product of quarrying at the local Mountsorrel site. As the haul roads became redundant, they were excavated and used as free-draining fill for the railway embankment stabilisation works.

The granite scalpins were also used beneath the earth embankment dam's reinforced concrete structure. This was to provide a drainage path, dissipating uplift pressures, and also to provide a firm working surface and foundation.

Figure A3.12a   *Railway embankment stabilisation works at the Melton Mowbray Flood Alleviation Scheme*

Figure A3.12b   *Foundations for earth embankment using scalpins*

Primrose Wharf, London – C&D waste from site used to fill gabion baskets in habitat enhancement scheme

A concrete-clad slope on the tidal foreshore at Primrose Wharf on the River Thames in London was partially broken up to enable gabion baskets to be laid in and planted up with *Phragmites australis* (Norfolk or common reed) in an effort to improve the habitat. The concrete demolished from the original slope was used as fill for the gabion baskets. Geofabrics were used to line the gabions and as a filter separation between the bottom fill layer and the planting medium above. Finally, pre-planted pallets of reeds were laid on to the planting medium. The scheme covered an area of 154 m² (4.1 m × 37.5 m) and used 30 gabions.

(Cook, 2000)

Figure A3.13a   *Construction of new foreshore using C&D waste-filled gabions planted with reeds at Primrose Wharf (courtesy of the Environment Agency)*

Figure A3.13b   *Planting the reeds in the planting medium in the tops of the gabions at Primrose Wharf (courtesy of the Environment Agency)*

River Bure, Norfolk – tyres used in riverbank protection

The Environment Agency has used rubber tyres in bank protection along the River Bure. This was carried out as part of a trial along with other methods of bank protection and proved to be successful. However, it proved difficult to install the tyres underwater as horizontal connections could not be obtained and backfill material was lost as a result of the movement of the tyres under wave action. Latterly it has been observed that the tyres had partially filled with silt and had good reed growth on the backfill.

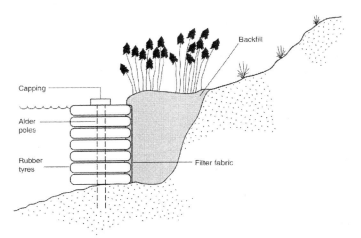

Figure A3.14a   *Tyre and post bank protection (schematic)*

Figure A3.14b   *Tyre and post bank protection*

Case Study A3.15

Pecos River, USA – tyre bales used for bulk fill/riverbank stabilisation

Technology is now available to compress and bale whole scrap tyres into cubes that can be used in sea defences and bank protection. Detailed pilot trials of these are being undertaken in the UK. Similar construction projects have already occurred in other countries.

Encore Systems reports that tyre bales were used in 1997 to stabilise the bank of Lake Carlsbad on the Pecos River in New Mexico, USA (see <www.tirebales.com/erosion_control.htm>. Wave action generated by passing water traffic had caused bank erosion. The project used around 250 000 recycled tyres. The river was drained and the tyre bales were placed in a trench about 1 m deep along the river's edge. They were placed on a concrete foundation containing steel reinforcing bars and were encapsulated in concrete. A concrete block retaining wall was constructed on top of the bales and fill was placed behind.

(Encore Systems Inc, 2001)

Case Study A3.16

Selby–Wistow–Cawood Barrier Bank, Yorkshire – colliery spoil used as bulk fill in river flood defences

The Selby–Wistow–Cawood Barrier Bank protects more than 2000 residential and other properties from flood overspill of the Yorkshire Ouse. At 5.5 km long it connects other defences at Selby and Cawood and crosses low-lying agricultural land. After constructing a trial embankment and conducting extensive testing to prove its suitability, colliery spoil was chosen as the principal fill material. This embankment averages 4 m in height and had a fill volume of 340 000 m^3. In economic terms, colliery spoil compared favourably with other materials available because of the proximity of the Selby coalfield spoil heaps.

(Kluth and Binnie & Partners, 1984)

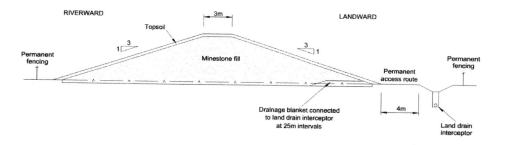

Figure A3.16   *Typical cross-section of the Selby-Wistow-Cawood Barrier Bank, Yorkshire*

Colliford Dam, Cornwall – china clay waste used as embankment fill for reservoir dam

Colliford Lake is South West Water's largest reservoir in Cornwall with a capacity of 29 100 million litres. The dam sited on Bodmin Moor in mid-Cornwall is an embankment 520 m long and 30 m high.

Both embankment and concrete gravity designs were considered and it was estimated that an embankment would be between 30 per cent and 40 per cent cheaper, largely because of the proximity of a spoil tip of china clay sand waste that could be used as fill. The local landscape adviser also considered that an embankment would be less visually intrusive and blend better with the surrounding moorland topography. Accordingly, the embankment dam ensured that both economic and environmental concerns were favourably addressed.

Although the china clay industry has years of experience in constructing tailings dams for its own settling lagoons, a trial embankment was built and tested before final designs were agreed. (For a full account see Johnston and Evans, 1985.)

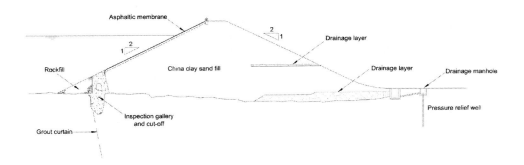

Figure A3.17   *Typical cross-section of the Colliford Dam, Cornwall*

(Johnston and Evans, 1985)

Slate used to be mined on Easdale Island, near Oban in Scotland, and in the adjacent village of Ellenabeich. The beaches and coastline in both locations are largely formed of slate spoil, which is now eroding rapidly, causing concern about the threat to village properties.

Figure A3.18a *Harbour seawall made of slate spoil at Ellenabeich, Isle of Seil, near Oban, Scotland*

Figure A3.18b *Erosion of old slate spoil wall at Ellenabeich, Argyll and Bute, Scotland*

Horsey Island, Essex – saltmarsh creation

Horsey Island lies within the inlet of Hamford Water, and is part of the Walton Backwaters. The island is strategically important in that it provides protection against wave action for the Backwaters, thereby reducing erosion. In 1998, the Environment Agency recharged 20 000 m³ of mud from Harwich Haven Authority's port development, between a shingle berm and the seawall. After nine months, considerable saltmarsh growth (*Salicornia* sp) had occurred over parts of the recharge area

The area was again recharged during January 2001, with the aim of raising the tidal height of the mud surface to facilitate the establishment of higher saltmarsh plants.

Suffolk Yacht Harbour, Levington, Suffolk – saltmarsh creation

Suffolk Yacht Harbour is situated on the east bank of the River Orwell, 6 km upriver of Felixstowe. Changes in the river have resulted in a loss of intertidal mud levels in the Orwell adjacent to the harbour. The harbour has always disposed of excavated material from the site locally, and has been keen to place the regularly available maintenance material on the adjacent foreshore. The dredged material is very fluid and pumped through pipes positioned within the marina to fixed pipes at the disposal area. The placement is within wattle hurdles or faggots (bundles of twigs stapled to the foreshore; see Figure A3.20). Sufficient material has remained to raise the tidal height on the foreshore to allow saltmarsh plants to colonise.

Figure A3.20 *Brushwood hurdles or faggots retaining dredged material on the foreshore near Suffolk Yacht Harbour, Levington*

Titchmarsh Marina, Walton on the Naze, Essex – saltmarsh creation

Titchmarsh Marina was constructed during the early 1970s to the south of the Walton Backwaters, close to Walton-on-the-Naze. The marina suffers regular deposition, as fine sediment settles from suspension and needs to be removed by frequent dredging. To allow for intertidal recharge of fluid dredged material, excavated earth was used to construct a bund on the west side of the marina. In 1998, sediments from the marina berths were dredged using the marina's grab dredger and the material placed into a barge before being broken down into a slurry that could be pumped into the recharge area.

Over two three- or four-month periods in 1998 and 1999, approximately 10 000 tonnes per year of mud was pumped into the bunded area (see Figure A3.21). In 2001, more mud was recharged, raising the tidal height further to allow saltmarsh plant development.

Figure A3.21   *Titchmarsh Marina bunded area of maintenance dredged material showing saltmarsh development*

Shotley Marina, Suffolk – shoreline replenishment with dredged material

Maintenance and capital dredged material from the Port of Felixstowe has been and continues to be used to repair a stretch of eroding coastline adjacent to Shotley Marina on the opposite bank. Figure A3.23 shows a digger (background) forming bunds from capital dredged gravel and maintenance dredged mud being pumped from the dredger *Sospan* into the area behind the bunds.

Figure A3.22   *Shoreline replenishment using dredged material at Shotley, Felixstowe*

PART OF SCHEDULE 3 OF SI 1994/1056 THE WASTE MANAGEMENT LICENSING REGULATIONS 1994

ACTIVITIES EXEMPT FROM WASTE MANAGEMENT LICENSING

(Parts relevant to this report – Paragraphs 7, 13, 15, 17, 24 and 25)

7.—(1) The spreading of any of the wastes listed in Table 2 on land which is used for agriculture.

(2) The spreading of any of the wastes listed in Part I of Table 2 on—
(a) operational land of a railway, light railway, internal drainage board or the National Rivers Authority; or
(b) land which is a forest, woodland, park, garden, verge, landscaped area, sports ground, recreation ground, churchyard or cemetery.

Table 2
PART I
Waste soil or compost.
Waste wood, bark or other plant matter.
PART II
Waste food, drink or materials used in or resulting from the preparation of food or drink.
Blood and gut contents from abattoirs.
Waste lime.
Lime sludge from cement manufacture or gas processing.
Waste gypsum.
Paper waste sludge, waste paper and de-inked paper pulp.
Dredgings from any inland waters.
Textile waste.
Septic tank sludge.
Sludge from biological treatment plants.
Waste hair and effluent treatment sludge from a tannery.

13.—(1) The manufacture from—
(a) waste which arises from demolition or construction work or tunnelling or other excavations; or
(b) waste which consists of ash, slag, clinker, rock, wood, bark, paper, straw or gypsum, of timber products, straw board, plasterboard, bricks, blocks, roadstone or aggregate.

(2) The manufacture of soil or soil substitutes from any of the wastes listed in sub-paragraph (1) above if—
(a) the manufacture is carried out at the place where either the waste is produced or the manufactured product is to be applied to land; and
(b) the total amount manufactured at that place on any day does not exceed 500 tonnes.

(3) The treatment of waste soil or rock which, when treated, is to be spread on land under paragraph 7 or 9, if—
(a) it is carried out at the place where the waste is produced or the treated product is to be spread; and
(b) the total amount treated at that place in any day does not exceed 100 tonnes.

(4) The storage of waste which is to be submitted to any of the activities mentioned in sub-paragraphs (1) to (3) above if—

(a) the waste is stored at the place where the activity is to be carried on; and

(b) the total quantity of waste stored at that place does not exceed—

(i) in the case of the manufacture of roadstone from road planings, 50,000 tonnes; and

(ii) in any other case, 20,000 tonnes.

15.—(1) The beneficial use of waste if—

(a) it is put to that use without further treatment; and

(b) that use of the waste does not involve its disposal.

17.—(1) The storage in a secure place on any premises of waste of a kind described in Table 4 below if—

(a) the total quantity of that kind of waste stored on those premises at any time does not exceed the quantity specified in that Table;

(b) the waste is to be reused, or used for the purposes of—

(i) an activity described in paragraph 11; or

(ii) any other recovery operation;

(c) each kind of waste listed in the Table stored on the premises is kept separately; and

(d) no waste is stored on the premises for longer than twelve months.

Kind of waste	Maximum total quantity
Waste paper or cardboard	15 000 tonnes
Waste textiles	1000 tonnes
Waste plastics	500 tonnes
Waste glass	5000 tonnes
Waste steel cans, aluminium cans or aluminium foil	500 tonnes
Waste food or drink cartons	500 tonnes
Waste articles which are to be used for construction work which are capable of being so used in their existing state	100 tonnes
Solvents (including solvents which are special waste)	5 cubic metres
Refrigerants and halons (including refrigerants and halons which are special waste)	18 tonnes
Tyres	1000 tyres

24.—(1) Crushing, grinding or other size reduction of waste bricks, tiles or concrete, under an authorisation granted under Part I of the 1990 Act, to the extent that it is or forms part of a process within paragraph (c) of Part B of Section 3.4 (other mineral processes) of Schedule 1 to the 1991 Regulations.

(2) Where any such crushing, grinding or other size reduction is carried on otherwise than at the place where the waste is produced, the exemption conferred by sub-paragraph (1) above only applies if those activities are carried on with a view to recovery or reuse of the waste.

(3) The storage, at the place where the process is carried on, of any such waste which is intended to be so crushed, ground or otherwise reduced in size, if the total quantity of such waste so stored at that place at any one time does not exceed 20,000 tonnes.

25.—(1) Subject to sub-paragraphs (2) to (4) below, the deposit of waste arising from dredging inland waters, or from clearing plant matter from inland waters, if either—

(a) the waste is deposited along the bank or towpath of the waters where the dredging or clearing takes place; or

(b) the waste is deposited along the bank or towpath of any inland waters so as to result in benefit to agriculture or ecological improvement.

(2) The total amount of waste deposited along the bank or towpath under sub-paragraph (1) above on any day must not exceed 50 tonnes for each metre of the bank or towpath along which it is deposited.

(3) Sub-paragraph (1) above does not apply to waste deposited in a container or lagoon.

(4) Sub-paragraph (1)(a) above only applies to an establishment or undertaking where the waste deposited is the establishment or undertaking's own waste.

(5) The treatment by screening or dewatering of such waste as is mentioned in sub-paragraph (1) above—

(a) on the bank or towpath of the waters where either the dredging or clearing takes place or the waste is to be deposited, prior to its being deposited in reliance upon the exemption conferred by the foregoing provisions of this paragraph;

(b) on the bank or towpath of the waters where the dredging or clearing takes place, or at a place where the waste is to be spread, prior to its being spread in reliance upon the exemption conferred by paragraph 7(1) or (2); or

(c) in the case of waste from dredging, on the bank or towpath of the waters where the dredging takes place, or at a place where the waste is to be spread, prior to its being spread in reliance upon the exemption conferred by paragraph 9(1).

Roger Maddrell	Halcrow
John Barritt	WRAP
Alan Brampton	HR Wallingford
Dave Brook	ODPM
Mark Buckley	Black & Veatch
Richard Clement	Ove Arup and Partners
Rod Collins	BRE
Tony Cosgrove	English Nature
Matt Crossman	Defra
Iain Cruickshank	SEPA
Jason Golder	The Crown Estate
Greg Guthrie	Posford Haskoning
Jennifer Hart	Halcrow
Brian Holland	Arun District Council
Elizabeth Holliday	CIRIA
Kerry Keirle	Welsh Assembly Government
George Lees	Scottish Natural Heritage
Patrick Mahon	Alan Baxter & Associates
Clare McCallan	Environment Agency
Stuart Meakins	Environment Agency
Fola Ogunyoye	Posford Haskoning
Anne Padfield	University of Greenwich
Phil Sanderson	Van Oord ACZ
Marianne Scott	CIRIA
Michael Wallis	HR Wallingford
Chris Worthy	Tarmac Recycling Ltd

References

AEA TECHNOLOGY (2002)
"Tyre stockpile survey carried out for the Environment Agency"
Unpub, Environment Council, London

ARMSTRONG, J M and PETERSON, R C (1978)
"Tire module systems in shore and harbour protection"
In: *Proc Am Soc Civ Engrs Waterways Div*, vol 104, no WW4, Nov, pp 357–374

BALDWIN, G, ADDIS, R, CLARK, J and ROSEVEAR, A (1997)
Use of industrial by-products in road construction – water quality effects
Report 167, CIRIA, London

BARBER, E C, JONES, C J, KNIGHT, P G K and MILES, M H (1972)
PFA utilisation
Central Electricity Generating Board

BIFFA WASTE SERVICES (2002)
Future perfect. An analysis of Britain's waste production and disposal account, with implications for industry and government for the next twenty years
Biffa Waste Services, High Wycombe, pp 91–94

BRANDON, T W (1989)
River engineering. Part II, structures and coastal defence works
Institution of Water and Environmental Management, London

BUILDING RESARCH ESTABLISHMENT (2000a)
Quality control. The production of recycled aggregates
BR392, Building Research Establishment, Garston

BUILDING RESEARCH ESTABLISHMENT (2000b)
Information Paper 7/00
<http://www.planning.odpm.gov.uk/material/index.htm>
[Accessed 02/06/03]

CHARLES, J A, TEDD, P, HUGHES, A K and LOVENBURY, H T (1996)
Investigating embankment dams. A guide to the identification and repair of defects
BR303, Building Research Establishment, Garston

CLARK, G (1992)
"Material production from aggregate quarries for Phase 2 and Phase 2a of the Morecombe Bay coastal defence works"
In: J-P Latham (ed) *Report on the proceedings of the seminar on armourstone, Queen Mary and Westfield College, London University*, pp 135–144

CLARKE, K (2000)
"The sustainability of reclaimed tyres for use in fluvial and marine construction projects"
Unpub MSc dissertation, University of Hertfordshire, Hatfield

COLLINS, R J (1998)
Recycled aggregates
DG433, Building Research Establishment, Garston

COLLINS, R J (1994)
Use of recycled aggregates in concrete
IP5/94, Building Research Establishment, Garston

COOK, J (2000)
"Primrose Wharf habitat improvements specification"
Unpub, Jonathan Cook Landscape Architect, 8 College Manor, 49–55 Farquhar Road,
London SE19 1SL

COVENTRY, S, WOOLVERIDGE, C and HILLIER, S (1999)
The reclaimed and recycled construction materials handbook
C513, CIRIA, London

CRABB, G and REID, M (2003)
Protocols for alternative materials in construction. Final report
Viridis, Crowthorne

DEPARTMENT OF THE ENVIRONMENT (1994)
*Environmental Protection Act 1990: Part II Waste management licensing. The Framework
Directive on Waste*
DoE Circular 11/94/Welsh Office Circular 26/94/Scottish Office Environment
Department Circular 10/94, HMSO, London

DEPARTMENT OF THE ENVIRONMENT, TRANSPORT AND THE REGIONS (1998)
*Statistics on arisings and use of mineral and construction wastes as aggregates: information
collection issues*
DETR, London

DHIR, R K, HENDERSON, N A and LIMBACHIYA, M C (1998)
Sustainable construction. Use of recycled concrete aggregate
Thomas Telford, London

DHIR, R K, LIMBACHIYA, M C and DYER, T D (2001)
Recycling and reuse of glass cullet
Thomas Telford, London

ELLIOTT, R, GHAZIREH, N and COLE, S (2003)
"Specifications and use in highways"
In: *Proc using secondary and recycled aggregates in construction. A SCiP roadshow, Hilton
Hotel, Newport, South Wales, 27 Nov 2003*
TRL, Crowthorne

EMMANUAL, B S (2002)
The use of recycled materials in coastal and river engineering
Unpub masters research, Univ of Liverpool

ENCORE SYSTEMS INC (2003)
Erosion control, Carslbad, New Mexico
<http://www.tirebaler.com/erosion_control.htm>
[Accessed 7/04/04]

ENVIRONMENT AGENCY (1992)
Assessment of erosion works
Data Sheet 3, Environment Agency, Bristol

ENVIRONMENT AGENCY (1999)
Waterway bank protection. A guide to erosion assessment and management
R&D publication 11, Version 1.0, Environment Agency, Bristol

ENVIRONMENT AGENCY (2001)
Flood defence investment strategy for England
Environment Agency, Bristol

ENVIRONMENT AGENCY (2002)
Solid residues from municipal waste incinerators in England and Wales
Environment Agency, Bristol

ENVIRONMENT AGENCY (2003)
<http://www.environment-agency.gov.uk/yourenv/eff/water>
[Accessed 01/07/03]

ENVIRONMENT AGENCY (in prep)
Reducing the risk of embankment failure under extreme conditions. Good practice review
Environment Agency, Bristol

ESCARAMEIA, M (1998)
Design manual on river and channel revetments
Thomas Telford, London

EUROPEAN COMMISSION (2000)
Management of construction and demolition wastes
<http://europa.eu.int/comm/enterprise/environment/index_home/waste_management/
constr_dem_waste_000404.pdf>
[Accessed 14/04/04]
DG ENV.E.3, Directorate E – Industry and Environment, European Commission,
Brussels

GILBERT, N (1996)
"Regulation of marine and coastal construction operations under FEPA (1985)"
In: J Taussik and J Mitchell (eds) *Partnership in coastal zone management*
Samara Publishing, Cardigan, pp 337–344

HALCROW MARITIME (2001)
Environment Agency's flood defence investment strategy for England. Final report
Halcrow Group, Swindon

HAM, R K and BOYLE, W (1990)
"Research reveals characteristics of ferrous foundry wastes"
Modern casting, Feb, pp 37–41

HAMILTON, W A H (1984)
"A colliery shale sea wall at Deal"
In: *Symp reclamation, treatment and utilization of coal mining wastes, Univ of Durham,
10–14 Sep*
Paper 45, NCB Minestone Executive

HAVANT BOROUGH COUNCIL
<http://www.havant.gov.uk/havant-2634>
[Accessed 19/09/03]

HEMPHILL, R W and BRAMLEY, M E (1989)
Protection of river and canal banks
Book 9, CIRIA, London/Butterworths, London

HOBBS, G (2000)
Reclamation and recycling of building materials: industry position report
IP7/2000, Building Research Establishment, Garston

HR WALLINGFORD (1977)
Colliery waste disposal on the Durham coast. Beach changes near Blackhall – a continuation study
Report EX765, HR Wallingford, Wallingford

HUMPHREYS, L (1996)
"A coastal morphology classification system for beaches on the Co. Durham coast
modified by the addition of colliery spoil"
In: J Taussik and J Mitchell (eds) *Partnership in coastal zone management, School of the
Environment, University of Sunderland*
Samara Publishing, Cardigan, pp 317–325

HUMPHRIES, B, COATES, T, WATKISS, M and HARRISON, D (1996)
Beach recharge materials – demand and resources
Report 154, CIRIA, London

HURLEY, J and McGRATH, C (2001)
Deconstruction and reuse of construction materials
BR418, Building Research Establishment, Garston

INSTITUTION OF CIVIL ENGINEERS (2002)
Design and practice guide. Coastal defence
Thomas Telford, London

JEWELL, R A and JONES, C J F P (1981)
"Reinforcement of clay soils and waste materials using grids"
In: *Proc ICSMFE, Stockholm*, vol 3, p 701

JOHNSTON, T A and EVANS, J D (1985)
"Colliford Dam sand waste embankment and asphaltic concrete membrane"
In: *Proc Instn Civ Engrs*, Part 1, no 78, Aug, pp 689–709

JOHNSON, T A, MILLMORE, J P, CHARLES, J A and TEDD, P (1999)
An engineering guide to the safety of embankment dams in the United Kingdom, 2nd edn
BR363, Building Research Establishment, Garston

JONES, C J F P (1985)
Earth reinforcement and soil structures
Advanced series in geotechnical engineering, Butterworths, London

JULIEN, P Y (2002)
River mechanics
Cambridge University Press, Cambridge

KLUTH, D J and BINNIE & PARTNERS (1984)
"Use of Minespoil in the Selby-Wistow-Cawood Barrier Bank"
In: *Symp reclamation, treatment and utilization of coal mining wastes, Univ of Durham, 10–14 Sep*
Paper 46, NCB Minestone Executive

LAGASSE, P F, ZEVENBERGEN, L W, SCHALL, J D and CLOPPER, P E (2001)
Bridge scour and stream instability countermeasures
Hydraulic Engineering Circular no 23, FHWA NHI 01-003, Federal Highway
Administration, US Department of Transportation, Washington DC
See <http://199.79.179.19/OLPFiles/FHWA/010592.pdf>

LATHAM, J P (ed) (1998)
"Assessment and specification of armourstone quality: from CIRIA/CUR (1991) to CEN (2000)"
In: J-P Latham (ed) *Advances in aggregates and armourstone evaluation*
Engineering Geology Special Publication 13, Geological Society, London, pp 65–85

LAWSON, N, DOUGLAS, I, GARVIN, S, McGRATH, C, MANNING, D and
VETTERLEIN, J (2001)
"Recycling construction and demolition wastes – a UK perspective"
J envl management and health, vol 12, no 2, pp 146–157

LEROUEIL, S, MAGNAN, J P and TAVENAS, F (1990)
Embankments on soft clays
Ellis Horwood, London

MASTERS, N (2001)
Sustainable use of new and recycled materials in coastal and fluvial construction: A guidance manual
Thomas Telford Publishing

MASTERS, N (2002)
"Crosby Marine Lake to Formby Point coastal strategy study"
Unpub, EX4496, HR Wallingford, Wallingford

MAY, R W P, ACKERS, J C and KIRBY, A M (2002)
Manual on scour at bridges and other hydraulic structures
C551, CIRIA, London

McCONNELL, K (1998)
Revetment systems against wave attack – a design manual
Thomas Telford, London

McDERMOTT, C (1987)
"Marine durability of mine waste"
Unpub BSc dissertation, Sunderland Polytechnic

MOTYKA, J M and WELSBY, J (1983)
A review of novel shore protection methods. Vol 1 Use of scrap tyres.
Report IT 149, HR Wallingford, Wallingford

OVE ARUP (2001)
North Wales slate tips – a sustainable source of secondary aggregates? (Summary)
National Assembly for Wales, Cardiff
<http://www.wales.gov.uk/subiplanning/content/minerals/secagg-e.htm>
[Accessed 10/10/03]

NORMAN, M J, HOUGHTON, J R, and JOHNSTON, J W (2001)
Hydraulic design of highway culverts
Hydraulic Design series no 5, FHWA-NHI-01-020, Federal Highway Administration, Washington DC

OFFICE OF THE DEPUTY PRIME MINISTER (2000)
Construction and demolition waste survey: England and Wales 1999/2000
Office of the Deputy Prime Minister, London

OFFICE OF THE DEPUTY PRIME MINISTER (2001)
Survey of arisings and use of construction and demolition waste in England and Wales 2001
<http://www.odpm.gov.uk/stellent/groups/odpm_planning/documents/page/odpm_plan_606009.hcsp>
[Accessed 02/06/03]

PERRY, J, PEDLEY, M and REID, M (2003)
Infrastructure embankments – condition appraisal and remedial treatment, 2nd edn
C592, CIRIA, London

PILARCZYK, K W (ed) (1988)
Dikes and revetments. Design, maintenance and safety assessment
AA Balkema, Rotterdam

REID, J M (2003)
The Aggregate Information Service. Final report
Project Report UPR/VR/013/03, Viridis, Crowthorne

SCHIERECK, J G (2001)
Introduction to bed, bank and shore protection
Delft University Press, Delft, The Netherlands

SIMM, J D (ed) (1991)
Manual on the use of rock in coastal and shoreline engineering
Special Publication 83, CIRIA, London/R 154, CUR, Gouda

SIMM, J D (ed) (1996)
Beach management manual
Report 153, CIRIA, London

SIMM, J D and MASTERS, N (2003)
Whole life costs and project procurement in port, coastal and fluvial engineering
Thomas Telford, London

SMITH, M R (ed) (1999)
Stone: building stone, rock fill and armourstone in construction
Engineering Geology Special Publication 16, Geological Society, London

SMITH, R A, KERSEY, J R and GRIFFITHS, P J (2002)
The construction industry mass balance: resource use, wastes and emissions
Report VR4, Viridis, Crowthorne

STUBBS, A (1998)
Environmental law for the construction industry: Mason's guide
Thomas Telford, London

UNITED STATES BUREAU OF RECLAMATION (1998)
The earth manual, 2nd edn
Part 1, Earth Science and Research Laboratory, US Department of the Interior, Washington DC

WASTE RESOURCES ACTION PROGRAMME (WRAP) (2003)
Classification of recycled and secondary aggregates as waste
News article, WRAP website <http://www.wrap.org.uk>
[Accessed 08/09/2003]

WATSON, C C, BIEDENHAM, S D and SCOTT, H S (1999)
Channel rehabilitation: processes, design, and implementation
US Army Engineer Research and Development Center, Vicksburg, Mississippi

WINTER, M G (1999)
"The use of spent oil shale in earthwork construction"
Geoenvironmental engineering, pp 565–574

WINTER, M G and HENDERSON C (2001)
Recycled aggregates in Scotland
Scottish Executive Central Research Unit, Edinburgh
<http://www.scotland.gov.uk/cru/resfinds/drf103-00.asp>
[Accessed 03/09/2003]

BRITISH AND EUROPEAN STANDARDS

BS

BS 882:1992 *Concrete aggregates from natural sources*

BS 3797:1990 *Specification for lightweight aggregate for masonry units and structural concrete*

BS 6349-1:2000 *Maritime structures. General criteria*

BS 6349:Part 2:1988 *Code of practice for maritime structures. Design of quay walls, jetties and dolphins*

BS 6349:Part 3:1988 *Code of practice for maritime structures. Design of dry docks, locks, slipways and shipbuilding berths, shiplifts and dock and lock gates*

BS 6349:Part4:1994 *Code of practice for maritime structures. Design of fendering and mooring systems*

BS 6349:Part 5:1991 *Code of practice for maritime structures. Code of practice for dredging and land reclamation*

BS 6349:Part 6:1989 *Code of practice for maritime structures. Design of inshore moorings and floating structures*

BS 6349:Part 7:1991 *Guide to the design and construction of breakwaters*

BS 6543:1985 *Guide to the use of industrial by-products and waste materials in building and civil engineering*

BS 7533:Parts 1–11:1997–2000 *Pavements constructed with clay, natural stone or concrete pavers*

BS 8002:Part 2:1994 *Code of practice for earth retaining structures*

BS EN

BS EN 12620:2002 *Aggregates for concrete*

BS EN 13043:2002 *Aggregates for bituminous mixtures and surface treatments for roads, airfields and other trafficked areas*

BS EN 13055-1:2002 *Lightweight aggregates. Part 1: Lightweight aggregates for concrete, mortar and grout*

BS EN 13055-2 [draft for consultation 00/107920 DC] *Lightweight aggregates. Part 2: Lightweight aggregates for bituminous mixtures and surface treatments and for unbound and bound applications excluding concrete, mortar and grout*

BS EN 13139:2002 *Aggregates for mortar*

BS EN 13242:2002 *Aggregates for unbound and hydraulically bound materials for use in civil engineering work and road construction*

BS EN 13383-1:2002 *Armourstone. Part 1: Specification*

BS EN 13383-2:2002 *Armourstone. Part 2: Test methods*

BS EN 13450:2002 *Aggregates for railway ballast*

National guidance documents

PD 6682-1 *Aggregates for concrete*

PD 6682-2 *Aggregates for asphalt and chippings*

PD 6682-3 *Aggregates for mortar*

PD 6682-4 *Lightweight aggregates for concrete and mortar*

PD 6682-5 *Lightweight aggregates for other uses*

PD 6682-6 *Aggregates for unbound and hydraulically bound materials*

PD 6682-7 *Aggregates for armourstone*

PD 6682-8 *Aggregates for railway ballast*

PD 6682-9 *Test methods for aggregates*

CONVENTIONS

The London Convention 1972. Convention on the Prevention of Marine Pollution by Dumping of Wastes and Other Matter

The Convention for the Protection of the Marine Environment of the North-East Atlantic 1998

Ramsar Convention 1971. The Convention on Wetlands

EUROPEAN DIRECTIVES

Assessment of the Effects of Certain Public and Private Projects on the Environment Directive 97/11/EC

Dangerous Substances Directive (76/464/EEC)

Directive on End of Life Vehicles (2000/53/EC)

EC Framework Directive on Waste (75/442/EEC as amended by 91/156/EEC)

Groundwater Directive (80/68/EEC)

Habitats and Species Directive (92/43/EEC)

Landfill Directive 1999/31/EC

Nitrates Directive (91/676/EC)

Water Framework Directive (2000/60/EC)

UK NATIONAL LEGISLATION

Coast Protection Act 1949

Control of Pollution (Amendment) Act 1989

Environment Act 1995

Environmental Protection Act 1990

Flood Prevention (Scotland) Act 1961

Food and Environmental Protection Act 1985 (Pt II)

Groundwater Regulations 1998

Land Drainage Act 1991

Waste Management Licensing Regulations 1994

Water Act 1989

Water Resources Act 1991

Wildlife and Countryside Act 1981

Further information

AggRegain

This sustainable aggregates information service run by WRAP (see below) was launched in February 2003. It is designed to assist anyone interested in specifying, purchasing or supplying secondary or recycled aggregates. The information provided comes from a wide variety of sources from within the aggregates and waste management sectors. The service will be updated and extended regularly, and anyone wishing to make a contribution or providing advice for the service should contact AggRegain directly.

AggRegain
WRAP
The Old Academy
21 Horse Fair
Banbury, Oxon OX16 0AH
Freephone helpline: 0808 100 2040
Email: helpline@wrap.org.uk
Website: <http://www.aggregain.org.uk>

British Ceramic Confederation
Federation House
Station Road
Stoke-on-Trent ST4 2SA
Tel: +44 (0)1782 744631
Fax: +44 (0)1782 744102
Email: bcc@ceramfed.co.uk
Website: <http:// www.ceramfed.co.uk>

Building Research Establishment
Garston
Watford WD25 9XX
Tel: +44 (0)1923 664000
Email: enquiries@bre.co.uk

BRE Scotland
Kelvin Road
East Kilbride
Glasgow G75 0RZ
Tel: +44 (0)1355 576200
Email: eastkilbride@bre.co.uk
Website: <http://www.bre.co.uk>

Centre for Environment, Fisheries and Aquaculture Science (CEFAS)
CEFAS Lowestoft Laboratory
Pakefield Road
Lowestoft
Suffolk NR33 0HT
Tel: +44 (0)1502 562244
Fax: +44 (0)1502 513865

CEFAS Burnham Laboratory
Remembrance Avenue
Burnham-on-Crouch
Essex CM0 8HA
Tel: +44 (0)1621 787200
Fax: +44 (0)1621 784989

CEFAS Weymouth Laboratory
The Nothe, Barrack Road
Weymouth
Dorset DT4 8UB
Tel: +44 (0)1305 206600
Fax: +44 (0)1305 206601
Website: <http://www.cefas.co.uk>

Environment Agency

Offices throughout the UK. *Call the general enquiries number to be put through to a local office*
General enquiries: 0845 9333111
Email: enquiries@environment-agency.gov.uk
Website: <http://www.environment-agency.gov.uk>

Encore Systems

PO Box 247
585 NW 3rd Street
Cohasset
MN 55721, USA
Tel: (218) 328-0023
Fax: (218) 328-0024
Email: baler@northernnet.com
Website: <http://www.tirebaler.com>

Etra (European Tyre Recycling Association)

Etra is the only European organisation devoted exclusively to tyre and rubber
recycling. It now has more than 250 members in 46 countries in Europe and around
the world. The membership includes policy- and decision-makers, collectors,
retreaders, manufacturers of recycling machinery, research bodies and developers,
among others.

Tel: (33) 1 45 00 97 77
Fax: (33) 1 45 00 83 47
Email: ETRA@euronet.be
Website: <http://www.etra.eu.com>

Environmental Data Interactive Exchange (EDIE)

EDIE is a free, personalised, interactive news, information and communications service
for water, waste and environmental professionals around the world. With comprehensive independent coverage, powerful search facilities, email alerts and discussion
forums, EDIE provides a one-stop-shop for the exchange of specialised information on
the Web.

This service is provided by:
Faversham House Group Ltd
232a Addington Road
South Croydon
Surrey CR2 8LE
Tel: +44 (0)20 8651 7100
Website: <http://www.edie.net>

Iron and Steel Statistics Bureau
ISSB Limited
1 Carlton House Terrace
London SW1Y 5DB
Tel: +44 (0)20 7343 3900
Fax: +44 (0)20 7343 3901
Email: info@issb.co.uk
Website: <http://www.issb.co.uk>

FIRST (Foundry Recycling Starts Today)
Elizabeth Olenbush, Executive Director
PO Box 333, Mill River
MA 01244 USA
Tel: (413) 229-2466
Fax: (413) 229-6091
Email: office@foundryrecycling.org
Email: inquiry@foundryrecycling.org
Website: <http://www.foundryrecycling.org>

Materials Information Exchange
Website: <http://www.salvomie.co.uk>

McAlpine Slate Limited
Penrhyn Quarry
Bethesda, Bangor
Gwynedd LL57 4YG
Tel: +44 (0)1248-600656
Fax: +44 (0)1248-601171
Email: slate@alfred-mcalpine.com
Website: <http://www.amslate.com>

Quarry Products Association (QPA)
156 Buckingham Palace Road
London SW1W 9TR
Tel: +44 (0)20 7730 8194
Fax: +44 (0)20 7730 4355
Email: info@qpa.org
Website: <http://www.qpa.org>

Used Tyre Working Group
The Group was formed in 1995 to act as a link between industry and government on used tyre recovery issues. It is formed of the directors of the four tyre trade associations, together with industry representatives and officials from the Environment Agency and the Department of Trade and Industry.

Website: <http://www.tyredisposal.co.uk>

TRL Limited (Transport Research Laboratory)
Old Wokingham Road
Crowthorne
Berkshire RG45 6AU
Tel: +44 (0)1344 773131
Busi tel: +44 (0)1344 770007
Fax: +44 (0)1344 770880
Email: enquiries@trl.co.uk
Website: <http://www.trl.co.uk>